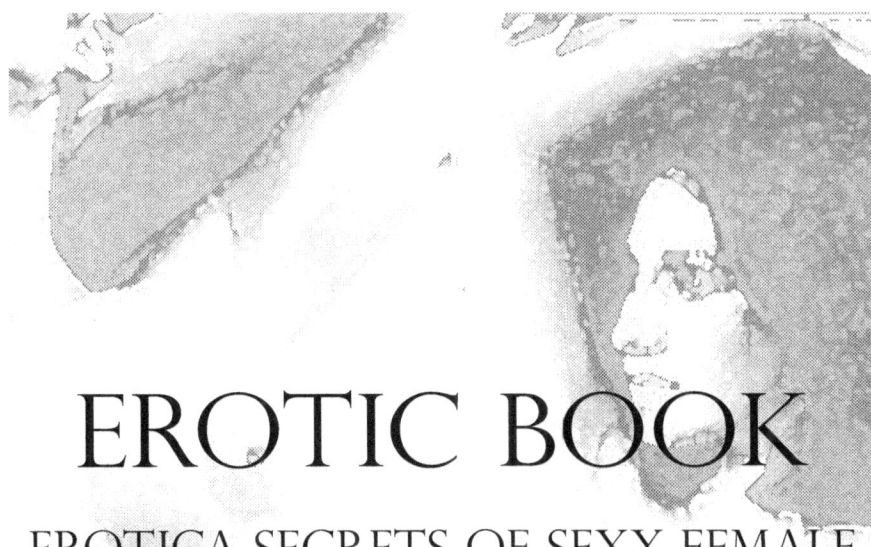

EROTIC BOOK

EROTICA SECRETS OF SEXY FEMALE BODIES FOR MEN AND WOMEN

SHELDON FILGER

Bloomington, IN  Milton Keynes, UK

authorHOUSE

AuthorHouse™
1663 Liberty Drive, Suite 200
Bloomington, IN 47403
www.authorhouse.com
Phone: 1-800-839-8640

AuthorHouse™ UK Ltd.
500 Avebury Boulevard
Central Milton Keynes, MK9 2BE
www.authorhouse.co.uk
Phone: 08001974150

This book is a work of non-fiction. Unless otherwise noted, the author and the publisher make no explicit guarantees as to the accuracy of the information contained in this book and in some cases, names of people and places have been altered to protect their privacy.

First published by AuthorHouse 4/4/2006

ISBN: 1-4259-2917-6 (sc)

Printed in the United States of America
Bloomington, Indiana

This book is printed on acid-free paper.

Cover Photograph by Sheldon Filger

Cover Layout by Jayne Cremasco

TABLE OF CONTENTS

INTRODUCTION

I have a unique perspective when it comes to an appreciation of the erotic nature of the female body: I photograph beautiful nude women.

While living in New York City, I experienced the devastation inflicted by terrorists on September 11, 2001. Like many New Yorkers, I sought ways and means of eradicating the melancholy gloom that overwhelmed the city's inhabitants. For myself, the best antidote proved to be an intensive immersion in beauty. This led me to a new calling, as a fine art photographer, specializing in the female nude figure.

During the past several years, I have photographed the bodies of numerous women. Each woman who has modeled for me is as different from the others as she is individualistic. They all have exhibited many distinctions of character, temperament, tonality, style, outlook and, of course, physicality. However, a common theme did emerge among my photographic subjects, the vast majority of whom had never posed nude before. Many of my subjects, prior to the photo shoot, had "issues" with their bodies. They either thought they were too heavy or too thin; their breasts were too large or microscopic. Because they perceived that the external world did not view them as sensuous and beautiful, that is how they came to view themselves. In the process, they constructed their own spiritual prison, inhibiting their ability to appreciate their own erotic nature. However, when they observed the photographic images that I created from our shoots, their universal reaction was one of jubilation. It was as though, for the first time in their lives, they could see and internalize how sensual and beautiful their bodies really were. Many women who have posed for me have remarked that this new awareness brought real empowerment to their lives and uplifted their self-esteem.

This book is intended for both women, and the men who love them. Based on my observations, I will share my ideas on what defines a sexy female body, as well as my conviction that almost every women is potentially a goddess of immense erotic power. It is my hope that both men and women will experience more pleasure in life through a greater awareness of the truly miraculous erotic creation that is the body of a woman.

ANATOMY OF AN EROTIC WOMAN'S BODY

Like most human beings, I have my own biases. I believe that the female body is intensely erotic, while that of the male is merely utilitarian. Obviously, many people will disagree with me. For example, there are men and women who view the male form as sensuous. However, what is truly important for our discussion is why I feel the way I do about the female form. A good place to start this conversation is in my photographic studio.

When a new model presents herself in my studio for the first time in a state of undress, I usually begin with a careful visual observation of her body. Her back is my starting point, and there is a reason. The back of a woman, extending from her head to her buttocks, is perhaps the most aesthetically underrated part of the female body. The site of an exposed back of a woman can be truly breathtaking.

What is it about the female back that I find so visually interesting? For one thing, it is far more aesthetically complex than that of her male counterpart. Due to the way the female torso is constructed, the structure of a woman's back has more intricacy to it. There is a delicate quality to the softness of a woman's shoulders, protrusion of shoulder blades and curvature of the spine. With the natural movement of arms and legs, the back of a woman becomes transformed into a visual symphony. The movement of bones and muscles creates dimples and crevices that magically appear, fleetingly, then disappear. I am always in awe as I observe the pure sensual delight that is projected from the backs of women.

Not surprisingly, many women are unaware of the visual splendor that lies just behind them. A good visual exercise would be to attend a cocktail ball, and observe the women wearing evening gowns with large parts of their backs exposed. There is a reason why barebacked dresses are a stylistically exciting fashion statement. While the thought of a man with an exposed back at a formal dance party would be anathema to most human beings, when it comes to women, an aesthetic triumph is assured. Praise be to barebacked women!

After keen observation of my subject's back, I then look at her buttocks, which again display a delicate character that differentiates them from those possessed by the opposite sex. There is a softness to a woman's buttocks that is both subtle yet sensual. Some believe that their shape replicates that of a full-bosomed woman with deep cleavage. From a structural standpoint, buttocks provide an element of sensual curvature to a woman's back that contrasts in a mysterious yet powerful manner with the curvature of breasts on the front of her body. A woman's bare back and buttocks combined present a very powerful erotic image. This is reflected in the fascination that artists and photographers have had for centuries with that aspect of the female anatomy.

While still viewing the rear side of my subject, I look with fascination at the back of her head. We all recognize that the front of the head is more visually intricate, with the complex and powerful presence of facial features. However, the rear portion of the head, in a more enigmatic way, always conveys something tangible about the nature of a woman. There is the interplay of shoulder length blond hair, or red hair tied in a bun, or a very short-haired brunette leaving the whiteness of an exposed neck juxtaposed with the darkness covering the top of her head. These examples merely hint at the infinite range of possibilities that challenge and titillate our visual senses.

Before viewing the frontal aspect of my subject's body, I have her turn around so I can observe her in profile. The shapes and curvatures exposed in a sideways view of a female figure are vastly more complex and interesting than that of the male counterpart. My eyes start at the top of her head, catching a glance of eye, the pinkish tint of a cheek, slope of a nose leading to sensuous, closed lips. With my keen and curious eyes I explore the unique and complicated curves and twists of the ear, often adorned with an ornate earring. I peer at her neck, noticing the protrusion of tendons navigating from the ear to the breastbone. Her arm and shoulder is an enigma, appearing both strong and fragile. The silhouette of the breast, sharply defined by an erect nipple, presents a statuesque proclamation of her feminine sensuality. I look studiously at the subject's diaphragm, heaving majestically with each breath of life. The bones from her ribcage protrude gently against the soft skin, and then I gaze at the side of her buttocks, the thigh and calf of her leg, all pulsating with energy.

When men and women think of the female form, they typically do so as strictly a frontally viewed composition. No doubt, as we shall be observing in later chapters, the frontal female nude figure is awesome in its raw sensuality. However, what I hope I have demonstrated is that all aspects of a woman's body are beautifully erotic. Though individual parts of the female form might be singularly beautiful, the ultimate visual miracle of the female anatomy is that its beauty is comprised of more than the sum of its parts. Thinking of the physicality of women in terms of a collection of body parts is not only demeaning, it denies both women and the men that love them the full measure of the sensual beauty of the complete and transcending female body.

We are still in my studio, and having completed my rear and profile observations, I am now prepared to carefully look over her frontal aspect. Everything I have observed before now comes together in aesthetic completeness. I again start with the head, appreciating the hairstyle of my subject, an expression of her uniqueness and individuality. I make eye contact, recognizing that a woman's eyes are a window to her soul. Her nose and mouth catch my attention, as the serious expression on her face is momentarily transformed when she cracks a brief smile, exposing a row of glistening white teeth.

I have viewed her neck during my profile observations. From a frontal view, it appears even more complex. Her collarbone protrudes sharply, above the breastbone, contrasting with the dimpled roundness of her shoulders.

Typically, breasts are viewed as the dominating terrain feature of the front portion of the female body. However, I think breasts are much more interesting visually when they are seen as an integral part of a complex torso composition, which includes collarbone, navel, abdomen and hips. Before I look intensely at the breasts, I like to observe the entire front torso aspect, and understand how the different anatomical components relate to each other visually. Only then do I study my subject's breasts, knowing I do so in a proper aesthetic context.

When I look at women's breasts, I am least interested in size, and will explain why in a later chapter. What truly intrigues me about breasts are their shape and texture. I view breasts the same way that I look upon a woman's face: a unique visual reflection on the character and personality of the woman. That is why I believe all women possess beautiful breasts .I never compare size; what I look at is how complex and interesting the curvature of the breasts are, their softness, and the shape and texture of the nipple.

The fascination with breasts is how their shape can be radically transformed through body movement. Gentle walking motion by women stimulates asymmetrical movement of their bosoms, as though the breasts are a duet embraced in a sizzling Argentine tango, spewing forth volatile erotic energy. Often, I will ask my subject to move her arms and stretch them upwards. This posture stretches the breasts, completely transforming their shape and texture. What is most visually exciting about breasts is that they are forever changing, tantalizing us with their visual transformations.

As I lower my gaze, I note with studious fascination how the abdominal region, with the navel in the center, appears almost like a desert depression or valley surrounded by the ridges that form the diaphragm. No architect, no matter how creative and freethinking, could possibly conceive so awesome and mysteriously beautiful a structure.

Navels are always unique. They come in a myriad of shapes, presenting their own customized visual statement amidst the abdominal tapestry. Some are dimple shaped, seeming to be aroused. Others are highly inverted, leaving a severely defined well of sensuality. Like breasts, their distinct appearance on every woman is an indelible mark of her unique identity. Adding to the personalization of the navel, many contemporary women choose to insert rings, some of simple designs while others are far more intricate.

After completing my observation of the abdominal region, I look analytically at my subject's pelvis and hips. The pubic hair in its darkness clashes intensely with the tender pale tone of the skin in the lower abdominal region. This complex cascade of colors and textures forms an erotic composition that is both blatant and simultaneously mysterious.

The hips of women I always find sensuous and strong. They are typically sharply defined, their curvatures manifesting an earthy feminine quality. I then lower my gaze, feasting my eyes on the smooth, shaved skin of her legs, the soft roundness of her knees and the pronounced strength of her thighs. My visual observation concludes with a glance at my subject's feet. I always find the feet of women more elegant in appearance than those of the opposite gender.

What have we learned from this foray into my studio? I believe two things are readily apparent. In the first place, the physicality of women is, by its very nature, inherently erotic. Secondly, to fully appreciate the eroticism of the female body, we must understand her visually as a complete artistic creation, and not a random collection of various external body parts, with varying degrees of sensual interest. A face alone is far less sensual than it is when combined with the woman's torso. Her back becomes even more arousing with a hint of the buttocks also exposed. Breasts are more alluring when the navel is also visible.

When I have completed my observations, I understand my model's body as a holistic erotic composition. Typically, the women who pose for me don't have that recognition of their own bodies until they have viewed the photographs I have created of them. When they view my images, it is a powerful revelation for them. In a very real sense, it is as though they are recognizing their own bodies for the first time.

Popular culture has given men and women a highly distorted and disjointed impression of what defines an erotic female body. Through both programming and advertising, television has relayed a false anatomy lesson on the sensuous woman. More often than not, the bodies of women have been reduced to commodities. It is as though the female person does not exist, she is just a living justification for some objects comprising hairdos, hints of breast and ultra-thin waistlines. These objectified women are placed on pedestals as the paradigm of what men find sensual about women. The subliminal hint to women is that those who don't conform to this commercialized paradigm lack sexiness in their physical form.

Most women on the planet do not look like what is represented on television and other avenues of popular culture as the ideal erotic woman. For that reason, many women develop "issues" with their own bodies that reflect the cruel personal cost of our societal immaturity when it involves female bodily aesthetics. Men and women are both in need of liberation from these false stereotypes created by the purveyors of advertising and mass culture, especially in the electronic media.

I have my own views on what defines an erotic female body. She is tall and she is petite, and just as likely of average height. Her weight is roughly proportionate to her height, with many variations possible based on her unique body build and bone structure, as well as lifestyle. She has an expressive face, complex back, rounded buttocks, breasts and navel, hips and legs. Her skin tone is pale, well tanned, black or of olive complexion. Her hair is brown, blond, red, black or perhaps frosted, arrayed in a diversity of styles. In essence, the great majority of women inhabiting our planet possess erotic bodies. Eroticism in the female figure is not the emanation of some form of mythic perfection. It is, rather, their "imperfections," meaning the differences that make each and every woman's body unique, that infuse the female form with that magical sensual presence and energy that is the life force of all humanity.

The famous and influential American writer, Ralph Waldo Emerson, wrote that, "though we travel the world over to find the beautiful, we must carry it with us or we find it not."

What Emerson expressed remains a very insightful observation for women. Each and every woman should recognize how beautiful and supremely erotic her own body is. The man or partner in her life likewise should follow Emerson's advice. It is perhaps appropriate to add another quotation, from a woman's perspective: "Each body has its own art," wrote the American poet Gwendolyn Brooks.

As a fine art photographer, my eyes are trained to recognize the unique artistry that characterizes every woman's body, and comprehend the erotic dimensions of her aesthetic beauty. In this book, I will share my observations on what characterizes sensuality in the female form and what most men find erotic about the body of a woman. Perhaps most importantly, I will try to enlighten women on what they need to know and understand about their own physicality, and the immense erotic power that it manifests.

FACE TO FACE TALK

How easily we forget that the head is an inseparable part of the female body. They coexist, aesthetically, in a state of dynamic tension. Since women are viewed the vast majority of the time in a clothed state, it is their faces that are the most frequently exposed part of their bodies. One thing has always been clear to me, an erotic body has an erotic face, and vice versa.

Both men and women often fail to understand the sensual power and emotion projected by the faces of women. Yet, it is the energy and character of a woman's facial features that are the first expressions of an irresistibly powerful erotic body. Understanding what is sexy about a woman's face is a path that leads to a more complete recognition of her holistic erotic presence.

What do I find so enticing about a woman's face? To begin with, I believe that the facial features of women are typically far more interesting than those of men. In my experience, men's faces often project toughness, anxiety, harshness and aggressiveness. It is as though all the trials and tribulations of life are chiseled on the male facial mask. With women, on the other hand, I am more likely to find a reflection of beauty, grace, and compassion along with strength and determination. It is in the faces of women that we first encounter the notion of beauty.

The ancient Greeks were clear on where they believed the secret of feminine beauty lay. Being mathematical in their observations of the physical world, they deduced that the proportions of a woman's face defined her sensual attractiveness. These proportions were sectioned into three parts: hairline to eyebrow, eyebrow to upper lip and upper lip to chin. This notion held through much of history, becoming somewhat more pedantic in the segmentation of the face during the Renaissance. In European culture, through the period known as the Pre-Raphaelite Revolution, artists set the societal norms in establishing notions of beauty. Faces of women could appear with chiseled features, hair heavy or flowing like water and eyes dark and expressive or soft blue and timid. A woman's face could be a representation of the Goddess Venus as with the Italian Renaissance painter Botticelli, or the anonymous sensuous gaze and smile that Leonardo Da Vinci painted on the visage of his Mona Lisa.

In our contemporary world, the recognition of what is a beautiful feminine face has become infinitely more confused. The advent of the entertainment industry as the primary driver in defining what is beautiful has tended to reduce common perceptions to whatever resembles the "hot" female celebrity of the moment. The actress Katherine Hepburn once remarked that the public perception of her possession of beauty was based merely on, "the good fortune to have been born with a set of characteristics in the public vogue."

To fully appreciate what is beautiful-and erotic-about the faces of women, we must transcend the artificially contrived set of characteristics that are "in vogue for the moment." Rather than the banality of the entertainment media, we should trust our own senses, which never lie.

My innate sense of the erotic compels me to observe with both fascination and pleasure the beauty that emanates from the face of almost every woman I encounter. What are the characteristics that define for me a beautiful feminine face? If truth should be told, the list could be endless.

I have seen women with small, pear-shaped faces with dark eyes, short blond hair, trimmed eyebrows and an expressive mouth, speaking with authority and determination, projecting power and even a hint of anger, and thinking that here I have discovered an erotic woman.

In other situations, I have encountered women with long, some-what narrow faces, subdued eyes, olive skin, thick eyebrows and substantial, curly brown hair, who spoke softly and were low-key in temperament. I knew I was in the presence of women with intensely sensual physiques.

I have been with women whose faces had large cheekbones, long noses and large ears, with long brown hair of silky texture, who rarely smiled, but had eyes that sparkled, and manifested pure eroticism.

There have been the occasions I have seen other women, such as one reading a newspaper in a subway car, unaware of my observations. With utter fascination I watched her brow compress, as her eyes skimmed that day's edition of The Wall Street Journal, her concen-tration only broken briefly as her hand brushed strands of sensuous black hair that had fallen across her forehead. Watching the sheer intelligence that flowed from every pore on her face, I sensed both her towering intellect and sheer feminine beauty and erotic energy.

What is the secret to a woman's face being both beautiful and erotic? The answer is actually deceptively simple. For the woman, it's recog-nizing that the same facial features she peers at each morning in the bathroom mirror need to be seen as fresh and new on every occasion, and sparkling with feminine character and sensual beauty. For men, it is the simple act of looking a woman's face, studying it, and capturing all its inherent beauty without resistance.

The first aspect we notice when viewing the face of a woman is her hair. The great eighteenth-century English poet, Alexander Pope, observed with stunning insight how, "beauty draws us with a single hair."

Her coiffure is the shop window to the male libido. Hair that appears sensuous to a man also arouses heightened expectations of her erotic physicality. It helps define the face of a sensuous woman. Before a woman can internalize her own sense of possessing an erotic body, she must think of her hair being sexy-and treat it that way.

Hair grooming is critically important to maintaining a sensuous facial look. Irrespective of the cut, color and tint, a woman's hair should always look neat and clean, with a sheen that reflects brightly when exposed to light. A studious regime of hair grooming communicates two things to men and the world at large. One is your own enhanced beauty. The other is more subliminal: here is a sophisticated woman with high self-esteem. She thinks highly of herself and her looks. When men form those thoughts from the well-groomed hair of women, they are then enticed to look at their faces and study them with vigorous curiosity.

Hairstyles are perhaps less important than a clean, well-groomed look, as long as the hair is treated seriously and not as an after-thought. Women should feel free to experiment, understanding that the metamorphosis of their hair is a far easier and less risky way of altering their facial appearance than cosmetic surgery. Based on my observations, there are a number of ideas on hair that I would like to share.

I have found that women who have very sexy hair understand that its purpose is similar to that of a frame found on a masterpiece paint-ing hanging in a museum gallery. With an exquisite painting, the frame is exactly that-it establishes the "visual frame" of that which captures our visual interest. Its role lies in defining the visual space we are observing for its artistic content and to complement it. For those reasons, fine works of art hanging on the wall of a distinguished art gallery tend to have frames that are not garish. They have an elegant simplicity, so as not to distract from the painting, or overwhelm it.

A woman's hair should serve the same complementary role as that of the frame I described above. It should be interesting, not garish, and its primary aesthetic role must be to complement the face, and encourage interested eyes to explore all the delightful and intriguing visual details that lie within the space defined by the hair. It should not be so elaborate and "loud" that is diverts a man's attention away from a woman's face, instead of redirecting it to the content framed by the hair.

When women experiment with different hairstyles, they too often focus their attention on the hair as a stand-alone visual phenomenon. My recommendation is to observe with studious attention how their new hairstyle complements their facial features. They should do this in a variety of states: intense and subdued lighting, smiling with joy or frowning with frustration; displaying anxiety or calm; wearing business attire, casual clothing, an evening dress, nightgown and being in the nude.

Women should understand that their hair projects the sensuality of their faces, and not only when viewed from in front and in profile. How their hair looks from the rear is frequently neglected, yet very often that may be the first visual encounter a man has with her. When observing their hairdo from a back perspective through mirrors, women should recognize that the interrelationship between hair and an erotic body becomes both more subtle and yet more complicated.

When I view the back of a woman's head, I am always intrigued by a number of visual phenomena. Are her ears covered, or exposed with interesting earrings? Is the hair cut short, exposing the back of the neck and shoulders, or does the hair cover up those areas, reflecting light from the strands falling on the shoulder blades? Has she rolled her hair up in a sculpture-like bun, or tied it into an intriguing ponytail? Women should experiment with and explore the "rear perspective" of their hairstyle alternatives and how they interact visually with the back view of their figures (obviously requiring at least two mirrors) with gusto.

In summarizing our discussion on hair, I am imploring women to understand how their hair interacts with their faces and upper bodies, creating a range of visual impressions. Men have the delightful task of being visually observant of what is being unveiled for them. To help build that understanding, let's get real close, for some face to face talk.

In judging what is beautiful and sensuous about a face, we typically would look at the face of a supermodel on the cover of a fashion magazine, and point at that reference as the face of a sexy woman. Yet, no photograph can even approximate the sheer eroticism that flows from the face of a living, breathing and animated woman. There is one reason that underlines that observations. The faces of erotic women are in constant motion. A man gazing at a sensuous woman is not beholden to a freeze-frame of her face, but stricken with rapture by the symphony of facial motion a sensuous woman projects by the mere movement of eye lashes, cheekbones and lips, with the periodic wave of a hand to adjust a strand of hair falling on the brow and shoulders. A sexy face as the integral part of an erotic female body is one that is a canvas of constant motion.

Why is movement is so crucial to projecting beauty and sensuality from the facial features of a woman? There are two principal reasons, I would submit. Number one, a face that has movement is expressive, and that contributes to a look that is far more interesting. Interest translates into the first link in the chain of sexual attraction. Number two, a face that is static as though frozen in a photographic image, no matter how otherwise superficially beautiful it may first appear to the observing male, quickly becomes monotonous, because it lacks passion. The most important defining characteristic of a sensuous female face is that it projects intense passion. Eroticism flows from passion, which in turn can only arise from a face that is constantly in motion, creating a never-ending stream of messages defining the woman who animates the face being gazed upon.

As I mentioned before, an erotic body is far more than the accumulation of diverse body parts. The face of a woman is a microcosm of the erotic female body. More than just the eyes, nose, mouth, chin, cheeks and ears, it is the interaction of all these features. The sensuous women knows that her face projects her passion and presence in the world, and she expresses her passion about her life with confidence streaming from every inflexion of her eyes, contractions of her lips and movement of her cheeks.

In my experience, most women do not comprehend how erotically powerful their faces are to men. To build that understanding, women should take advantage of their morning visits to the bathroom mirror, and study their own facial features as though observing them for the first time.

Women should look into the reflection of their facial features as they undergo a rapid metamorphosis. Understand the impact of the subliminal messages they exude when showing a range of emotions. Tension in the facial muscles, tight lips, intense gazing from the eyes and uplifted eyebrows convey one set of impressions. Relaxing the facial muscles, smiling widely with teeth exposed, combined with a hint of shyness in the eyes, sends a very different set of signals to the opposite gender. Adding to the infinite range of possibilities created from facial manipulation are the additional aesthetic complements of make-up and beauty aids.

Make-up, used with flair and sophistication, can significantly augment the sensual power of the female face. The mistake I see many women make is to confuse make-up's essential purpose, and see it as a panacea in lieu of a complement. When make-up is used successfully, it enhances the natural appearance of the woman's visage rather than disguise it. When employed unsuccessfully, it overwhelms a woman's natural beauty, transforming her face into a caricature. Men of distinction are highly attracted to a natural look that has been artfully augmented by the deft employment of cosmetics and beauty aids. They are also turned off by excessive camouflage of the face by multiple layers of rouge and an over-abundance of lipstick and mascara.

Lips are the most intimate feature on a woman's face. As with other cosmetics, lipstick, used skillfully, can enhance their sensual character. As I suggested earlier with hairstyles, the erotic woman should invest her time in experimenting with different shades. While viewing her reflection in the mirror, she should judge her choice of lipstick not merely by staring straight at her own image. Assessing how the shade of lipstick complements the rest of her face, she should manipulate her facial muscles and lips, projecting a range of emotions. A shade that appears sensuous when smiling must also be sexy with subtlety when the face is animated by lively argument, or frowns with a melancholy feeling. Turn your head to the right and left; and observe if your lips still appear sensuous when your significant other or date can only view them in profile.

In Hollywood fantasies, a deep red lipstick is often portrayed as being a neon advertisement for a sensuous female. I beg to disagree with that stereotype, with one exception. Usually, "loud" lipstick shades send strong signals to observing men, but not of a quality that is complementary to a proud, sophisticated and erotic women. My preference in most situations is for a more subtle selection of lipstick color. The only circumstance I feel more aggressive lipstick shades do convey sophisticated sensuality is in low light situations, such as one may encounter at the cinema or theatre. In a nocturnal environment, lipstick that emphasizes the presence of a woman's lips can be a very powerful manifestation in a situation where otherwise the facial features cannot be easily discerned.

Eyes are the windows of the human soul. The interrelationship of eyes, eyebrows and eyelashes are an essential dynamic in determining the erotic character of a woman's face. As with hairstyles, lipstick and other cosmetics, women should explore different and subtle ways of using eyeliner, eye shadow and mascara. Many women, especially in middle age, approach these cosmetic aids as ways of capturing a youthful look. A far better way of emphasizing a woman's erotic essence is to make deft sue of these aids to project a look of sophistication and mystery. Intrigue is sexy, and no other aspect of a woman's face is so intense and bold in identifying her as sublime and intriguing.

The term "eye contact" is often used to describe the first contact between men and women involving communication of mutual sensual interest. A woman's eyes are very powerful attributes to her erotic nature, and knowing how to use them in a sensuous way is even more important than cosmetic aids. As with other features of the face, the erotic woman's eyes are in constant motion, creating a stream of powerful impressions to be absorbed by the male observer and his libido.

Alternating between direct eye contact, than lowering her gaze to show interest in other aspects of her male companion, than looking downward in contemplation of her own physique, in combination with alternative straight lips and sensuous smiles, is a highly effective technique for facilitating arousal. Eyes that are lively are far more interesting than those that are static, as though locked on a specific object Men notice eyes that convey an interesting feminine mind and body. Interest, in this context, correlates directly with the perception of a woman having an erotic body.

Now that we have explored the head of a woman, front and back, let us examine the issue of angles. It may seem peculiar to inject geometry into the sensual equation in the issue of facial sensuality. However, the topic relates directly to a point I have already stressed; facial movement and its linkage to building a perception of sensuality.

In face-to-face conversation, women have a tendency to keep a "straight face." While perfectly appropriate in a business or professional context, such rigid facial posture is actually antithetical to our goal of projecting an erotic gaze when encountering a man of interest. When seated across the table from a male companion, shifting the angle of your face is another form of dynamic movement. It should be done naturally and not with a jerking motion. However, done with subtlety, another form of subliminal projection of your innate erotic power is being accomplished.

To understand how shifting head posture can arouse more interest in your facial features, let us again visit your bathroom mirror. Imagine yourself as your partner or boyfriend, looking directly at you. The next step is to periodically tilt your head, just a few degrees, first to the right, than to the left. As you do that, smile briefly, gently move a shoulder towards your ear, and occasionally brush your chin and lips with your fingertips.

The simple movements and angling of the head, done very naturally, will heighten the visual interest that your male companion derives from your face. He will look at your facial features with greater intensity, and feel a heightened sense of arousal. How is this achieved? By the simple means that you have transformed an act of conversation or simple eye contact into a situation where your companion is compelled to concentrate on your face with greater intensity.

Having discussed the sensuous face, let us now look at the issue of props, specifically earrings and eyewear.

Ear piercing is an ancient form of body modification. Cultures from the dawn of civilization have incorporated earrings as one of the most prolific forms of jewelry. Though it is increasingly acceptable in Western culture for men to wear earrings, they are still viewed as quintessentially feminine ornamentation.

Earrings enhance the erotic appeal of women. In the first place, they juxtapose their own innate design with the natural complexity of the ear. This visual interaction is highly enticing to most men. Secondly, as with the selection of hairstyles, they are a powerful form of personalization by women. The myriad choices in ear jewelry available to the modern woman offer an infinite array of possibilities to further enhance the erotic appeal of her facial features.

Typically attached by a piercing in the earlobe, as well as other external parts of the ear (especially if multiple piercings are done), earrings can be fabricated with a variety of materials, including metal, glass and precious stones. The designs vary greatly and can include studs, plates and loops. They can be small and almost inconspicuous, or large and ornate.

As with hair, I find the sensuous face presents earrings that complement its features, rather than being a visual distraction. I highly recommend against heavy earrings, which can stretch and even damage the earlobe-an occurrence with does nothing to enhance sensual appeal. On the other hand, women should not be bashful about being bold and even adventurous with their earrings.

Often, I find women wearing exotic earrings, frequently purchased while on vacation to some dreamlike destination. Men are intrigued by such facial ornamentation. They are compelled to look at the earrings with admiration, simultaneously coming into visual contact with the intrinsically complex spiral curves of the ear. The impression is highly erotic, reinforced by the perception of innate sophistication such ear ornamentation conveys.

Some of the most sensuous earrings I have seen are loops fabricated from metal. Deceptively simple, when attached to the earlobes of women, they complement their faces in a very powerful and erotic manner. In particular, women with short hair, with their ears fully exposed, present a highly sexy demeanor to the world when wearing large looped earrings.

In our discussion on earrings, I have commented on how they enhance the sensuality of a woman's face. However, their erotic impact is not limited to the face alone. Women in a nude state can appear even more erotic by wearing their earrings. In other words, earrings also enhance the erotic appeal of the entire body. I consider earrings to be the most erotic form of female jewelry.

As with my other suggestions, women should experiment with a selection of earrings while viewing the bathroom mirror. Observe how they interact with your facial aesthetics while presenting a variety of emotional states. Explore how they complement different clothes, from dresses and business attire to casual outfits and nightgowns. Look at your earrings while naked, and comprehend how they add an element of visual ornamentation to all the wondrous aspects that make up your physical body.

We now come to the issue of eyewear. Women who are nearsighted, as well as those with excellent vision, have the same need for protection from intense sunlight. Sunglasses perform a valuable protective function, however, they are also a fashion statement with a sensual dimension. Women who wear sunglasses in appropriate situations should fully recognize that they add to her erotic appeal. Walking along a busy city street on a sunny day, virtually any woman appears sexier with shades on. This is irrespective of head shape or hairstyle, or her overall physicality.

We discussed earlier how eyes in their expressiveness and intensity are the windows to a woman's soul and psyche. Closing those windows in broad daylight opens another pathway, the strong hint of intrigue and mystery. When that mystery woman animates other parts of her face, especially lip movement, in lieu of eye contact, the impact on the male observer is intense. His libido is activated, and his interest in the woman is heightened. "Who is that woman of mystery that I find so incredibly attractive and sensuous?" he will ask himself.

When is it proper to wear sunglasses? I see shades as eyewear for the outdoors. They are not indoor props, and don't work with full nudity. However, women who are fortunate to have access to a legal topless beach should appreciate the sizzling combination of surf and sand, bare breasts and very dark sunglasses.

Women who were not born with perfect eyesight may wonder what is more advantageous for projecting a sensuous face, eyewear or contact lenses (or surgical correction of nearsightedness)?

For the most part, men prefer a natural face. However, not all women who are nearsighted are comfortable wearing contacts. They need not despair. Eyewear, especially with the vast assortment of designer frames available, can be, especially if well chosen, a complement to a sensuous face. Many frames are in themselves sexy in design, and act out an intriguing visual juxtaposition with the eyes and other facial features of a woman. Women who are nearsighted and have the ability to wear contact lenses, have the fascinating option of alternating between wearing contacts and eyeglasses, adding to the range of sensual perceptions men will form of their facial features.

My comments on eyewear pertain to being in a clothed state. In the nude, it is wise to abstain from wearing glasses. Otherwise, they form an asexual incongruity that obstructs the man's perception of an otherwise erotic female body.

In concluding my commentary on the role of a woman's visage in defining her physical body as erotic, I am reminded of the lines Christopher Marlowe wrote about Helen of Troy in *Faustus* to immortalize the power of a beautiful, sensuous woman's face to arouse passionate, erotic love from an enraptured man: "Was this the face that launched a thousand ships, and burnt the topless towers of Ilium?"

BREASTS OF ANGELS

"Thy two breasts are like two young roes that
are twins, which feed among the lilies"
Song of Solomon, Chapter 4, Verse 5

The fact that the Holy Bible contains an erotic ode to the breasts of a female beloved only hints at the epochal fascination and awe men have had for this anatomical feature of the female torso for millennia. There exists no other part of the female anatomy that evokes so much conversation, verbal lust, introspection and fantasization among men as the breasts of women. This connotes the almost surreal attention and focus men have regarding breasts.

The association of eroticism with female breasts transcends the realm of intimacy, and is a pivotal factor in marketing products to a male clientele. A non-too-subtle example is how car designers seek to imitate the rounded lines of breasts. During the 1950's, for example, it was an open secret that the bulging chrome bumpers Detroit designed into many of its automobiles were intended to suggest a buxom woman.

Breasts are clearly a potent aspect of the female anatomy when it comes to evoking sensuality in the presence of a man. Yet, there is a dark side to this phenomenon, which has created immense and unnecessary emotional suffering in women. It has to do entirely with the myth that the erotic beauty of a woman's body is solely correlated with the size of her breasts. However, nothing could be further from the truth

In opening our discussion of what defines a sensuous bosom, we must first kill the myth that mass and volume are the sole relevant determinants of breast aesthetics. In point of fact, that is the least important visual aspect of breasts.

Let us, for the sake of our discussion, have an imaginary contest for men. Photographs of two women are displayed to a group of heterosexual males. One image shows a female with a 36-inch bust line. The other woman's bust line is of modest dimensions. They both possess looks that otherwise would be considered attractive. The men are asked to vote on a simple selection: which of the two women do they wish to see topless? If you assumed that the overwhelming choice would be the woman with the "big tits," to use a colloquial expression, you have in all probability guessed correctly.

Continuing with our exercise, let us again ask our imaginary males the same question, however, we will add an additional piece of information for them to reflect on before rendering their selection. While the female who is "well-endowed" is an anonymous professional woman, the other woman with an inconspicuous bust line is a celebrity. She could be an actress, or the leading anchorwoman for the evening news, a prominent politician, maybe even the mayor or a senator. Now, who do you think the vast majority of men would prefer to see topless?

In our first scenario, men expressed the preference that they have been conditioned to render. In the second situation, additional information transforms the women in the perception of the male audience, and one becomes far more interesting, based on her own achievements. She is the one they want to view "topless." The essential point is that breasts are intensely erotic not because they are large or small, but due to being the shapely and intimate aspects of women who intrigue and fascinate men. It is not only celebrities who possess that power over the male libido. Any woman who is comfortable in her own body, and has created erotic interest through the sensuality of her facial features and personality, possesses a bosom that leaves men who are classy and distinguished salivating at the prospect of viewing it unveiled, in all its glory.

Lesson one about erotic breasts is that size means nothing. How much personal agony and emotional insecurity would be spared if this simple gospel were propagated throughout the world is incalculable.

According to the American Society for Aesthetic Plastic Surgery, in 2004 more than 334,000 women in the United States underwent breast augmentation procedures. How ironic it is that so many women, in a culture otherwise obsessed with slim female figures, endured the risks and discomfort of surgery, merely to stuff their bosoms with artificial fat!

Let us be blunt, the mass or size of the breast is a function of adipose tissue, otherwise known as fat. Saline or silicon gels implanted into breasts replicate fat content. Would women otherwise consider cosmetic surgery involving implants placed in their buttocks or hips or even their chins? Probably no rational woman would ever consider such an option. Yet, often in despair, millions of very intelligent and beautiful women have been stampeded by crass cultural stereotypes, supposedly about feminine beauty, to subject themselves to having their breasts sliced open by a surgeon's knife, then stuffed with an implanted foreign substance not meant for the human body. Often, such dire measures lead to tragic consequences, with legions of horror stories concerning botched operations and leaking implants. Such a price many women must pay, solely because of an illogical myth that fatty breasts are the mark of Venus. Yet, how many Renaissance artists painted a sensuous version of that seminal Greek goddess with titanic boobs? The answer is virtually none.

It would be most instructive for women and men to look anew at all the great works of art celebrating the female body, and glorifying its erotic character. The overwhelming majority of such paintings portray women with breasts that are, if anything, "small." It was a common practice for Renaissance painters to employ boys as their "female" models when painting the nude female figure, so convinced were they that a small bosom on the female figure represented the quintessence of feminine beauty and sensuality. Similarly, great fine art photographers of the female nude have consistently demonstrated a preference for models with bust lines of modest proportions.

My observations are not intended to disparage women who have large breasts. My point is that breast size, large or small, is only one aspect in a complex array of features that determines the bosom's aesthetic beauty in the eyes and minds of men. Among males, it is only the boorish and vulgar who are obsessed with gargantuan boobs, to the exclusion of every other aspect of a woman's remarkable body. What woman of quality should have to feel inadequate unless she undergoes a surgical operation merely to appease such an inane and vapid male? The answer is none.

One other point to consider, concerning breast size. It is estimated that approximately 60% of American women wear bras with an A or B cup. Having small or modestly dimensioned breasts is actually quite in fashion, at least according to Mother Nature.

From my perspective, what are most aesthetically intriguing about female breasts are their shapes, colors and textures. They are so different and unique on each woman, they are, in a sense, her second face. It is a mark of her individuality that seethes with eroticism, becoming the forbidden fruit that a man can only partake of if he is worthy of arousing her passionate interest in him. To briefly return to our earlier exercise involving the male preference on selecting a woman's breasts for exposure, my experience tells me that men who find a woman of great sensual interest are desperately seeking to view her forbidden "second face." They can feel within the deepest recesses of their subconscious that the breasts of a woman with a beautiful, sensual face and intelligent, lively mind will be of far greater erotic interest than those of another woman without the same alluring qualities, irrespective of how massive her bra size might be.

Shall we now delve into what is truly erotic about the female breast? I'm ready if you are.

Let's us first explore the diversity of shapes of breasts, for that is typically the first visual characteristic that men salivate over when first coming into visual contact with these "objects of desire." It is truly a miracle of creation that women of our human species possess bosoms with such a high degree of visual differentiation, especially in comparison with the many other species of mammals that walk upon our planet.

The anatomical structure of a woman's breast is the explanation for the range of configurations encountered when viewing the naked female torso. The Coopers ligaments sustain the primary anatomical support for the breasts. Additional support also is derived from the skin that covers the breasts and it is this support which determines the shape of the breasts

In my work photographing the bodies of women, I have observed breasts that are perfectly rounded, and others that bulge majestically. Some form a shape that is almost tubular and appear to spread apart, simulating an inverted V-shaped object. Others are broad, and can be referred to as "pigeon-breasted." While breasts tend to sag with age, larger breasts will droop earlier in a woman's life, and many men actually find this to be a highly erotic appearance. In other cases, I have seen large breasts that were very firm, forming a sensuous cleavage. I have also viewed women who were almost flat chested, creating a lithe and athletic visual impression. Other women had small breasts with a delicate, semi-circular shape.

There exist an infinite number of variations among the breast shapes I described above. With billions of women living in our world, there are bound to be. The shape, or to use an artistic term, the line of the breasts, defines its inner space, in much the same manner that I described in an earlier chapter in connection with hair framing a woman's facial features. That defined shape captures the man's imagination and arouses his interest in exploring the wondrous sensual aesthetics that are contained within that erotic configuration.

Things are even more complicated for the male libido, however, when it comes to viewing a woman's delectable breasts. For, it happens that the shape of breasts can be transformed, even with small body movements. The descriptions I have just shared of breast shapes only apply to a standing female body, viewed frontally.

Observe the torso of a woman in profile and things begin to change drastically. An intriguing hump is clearly in view, and forms a profile exposure, that line of curvature on the upper torso of the female body that is probably the most sensual curvature any man can observe. Even with small breasts it is clearly defined. Often, an erect nipple, outlined in profile, will add a further erotic dimension to the scenery. When the lights are dimmed, that curved profile becomes seductively quilted with shadows, contrasting with an alluring erect nipple. The many ways a man can observe the breast profile of his beloved serve up a smorgasbord of delectable sensuality.

When I photograph the torsos of nude women, one of my favorite perspectives is to view them obliquely. This enables me to peer at one breast frontally, while simultaneously seeing the other boob in profile. The juxtaposition between the two breasts of the same, sensuous woman is utterly awesome in the expression of unrestrained feminine erotic power. This leads me to another observation on the erotic aesthetics of the female breasts. Asymmetry is much more interesting than conformity. Just because a woman has two breasts, it does not mean they both have to march to the beat of the same drummer. As the French say, vive la differences!

A woman's genetic make-up seems to be cognizant of the heightened sensual quality of breasts that are not quite exactly the same. More often than not, women have one breast that is slightly larger than the other. The breasts may not be of exactly the same configuration. While one nipple is erect, the other may be temporarily spread out, as though in a state of relaxation. All this nonconformity of the breasts is visually breath-taking for men, as well as being highly stimulating for their libidos.

When a woman is rendering her male companion the high honor and privilege of observing her boobs, she should offer them up with as much diversity as possible. Perhaps it is time to again look at the mirror.

A woman trying to understand what men see as fascinating about their breasts should look at them straight ahead, then slowly twist her upper body to one side, transforming the frontal view of her breasts to a profile perspective. Study the curved line defining the breast, peer at the nipple, as it stands erect, then dim the lights, and comprehend how their appearance is radically altered. Now, twist slowly back to your original position, until one breast remains primarily in profile, while the other can be seen frontally. Contrast the two breasts in terms of shape, color and texture. That is exactly the process that occurs within the male psyche, almost entirely driven by an immensely powerful subconscious desire.

Breasts should be thoroughly enjoyed and relished for their unique visual sensuality, from a wide array of perspectives. When standing, women should be aware how movement of their arms can radically alter the configuration and overall appearance of their breasts, further tantalizing the male observer, as his libido erupts with volcanic intensity.

Stretching the arms upward forces ligaments and skin to lift the breasts, leading to a flatter appearance. With small breasts, they may blend into the torso, providing for a graceful and athletic appearance. With larger breasts, an entirely different circular line is formed, oozing sensual fascination. Women with sagging breasts will find that stretching the arms upwards, or even placing the hands behind the neck, will raise the bosom, displaying a firmer and more rounded appearance.

The above description relates to another important observation about erotic breasts. Interplay of arm movement and breast shape is another secret to delighting the male observer with an infinite menu of aesthetic sensuality. Women who desire to understand this dynamic better should keenly observe how the shape of their breasts is altered by bringing the arms up to shoulder length, then placing the hands behind the neck and then stretching both arms to the ceiling. Once the sensuous woman has become acquainted with her ability to transform the appearance of her breasts through arm movement, her next step is to add even more complexity to the visual equation. Try stretching one arm upwards, while keeping the other hand on your hip. Notice the entirely different configuration that characterizes your bosom. Imagine how tantalizing such a visual metamorphosis is to your partner. Oh, if only you could imagine!

As with our earlier discussion on the importance of movement in defining a sensuous female face, the identical principle holds true in connection with boobs. Most women's breasts are beautiful and erotic, but are not understood as such by their "owners." Many women have a tendency to think that their breasts should be viewed as static objects, as though she was merely a statue.

Erotic women are interesting and dynamic. They exhibit constant movement and transformation in their body language. Just as their eyes open and close seductively, and their lips alternate from mystery-evoking tightness to celestial wide-open smiles, their bosoms also should be "displayed" with similar alacrity. Don't just let your beloved view them as a couple of milk-producing glands suspended from your chest wall with monotonous regularity. Twist and turn so they bounce, alternate between frontal and side exposures, with as many variations in between as possible. Learn to naturally move your arms and hands while being observed bare-breasted, stretching and altering the appearance and aesthetic line of your breasts in every conceivable way.

The bottom-line message to women on breast shape and your God-given ability to engage in boob transformation: Go wild! And men, enjoy the ride!

Having discussed breast shape and size, let us now come to the core matter defining the aesthetic and erotic character of the female breast. In my view, color and texture are the most essential visual elements in determining the visual eroticism of the breast. Though one wouldn't think so based on all the folklore about breast size and shapes, the clash of colors and textures that occur on a woman's chest are perhaps the most erotic visual factors on her body.

In evaluating breasts and their color, we typically think of boobs replicating the skin color of the woman as observed elsewhere on her body. In reality, that is only partially true. As we know, when a woman is out in the sunshine receiving a tan, at least in most of the world's cultures, her breasts will remain covered. On a typical beach in the Western world (for the sake of this discussion, we will exclude "topless" beaches), women may wear skimpy bikini tops exposing a large portion of their breasts to the rays of the sun. Even so, the lower portion of the breasts, including the nipples, remains covered.

The result of tanning on women's bodies is that, when they are viewed bare breasted, clearly discernable tan lines are observed. Many women mistakenly assume that tan lines on breasts are unattractive to men. Some females resort to tanning salons and other solutions to dim the visual demarcations on their breasts formed by a combination of natural sunlight and socially imposed beach ware conformity.

I offer a vastly different perspective on tan lines and tits. They not only look terrific, they are inherently erotic. The clash of skin tones on the female breast brings out a strong element of visual complexity that catches the attention of the male observer, stimulating a strong flow of erotic interest. Tan lines on breasts are tantalizing, alluring, sensually powerful and artfully erotic accoutrements to the female body. The most intriguing portraits of beautiful nude women, whether created by painters or fine art photographers, will always reveal clearly defined and mesmerizing tan lines on breasts.

The combination of varying skin tones on breasts is endless. Even women with fair skin have them, revealing a seductively subtle transition in shades of pinkish-white as her partner's eyes survey her bosom with heart-racing interest. With darker-skinned women, a whole different pattern of skin tones is revealed. Perhaps the most visually erotic skin tones are found on the breasts of well-tanned Caucasian women. One example comes to mind, as it certainly left an indelible visual impression. I recall a tall, blond woman, skin nicely tanned, including the upper half of her breasts. In startling contrast, the lower half of her modestly sized breasts were a delicate shade of white, synonymous with Nordic women. To add to the visual complexity, her fair-skinned tan lines enveloped a large, dark and powerfully shaped nipple. I can't recall any abstract work of art that was so overwhelming to my senses as the cascade of colors and textures on that woman's bosom.

Men and woman need to smell the roses when observing breasts. To often, both the women who possess these wonderful objects and the men who are tantalized by them will give them the "once over." Breasts should be studied with one's eyes as though they are a work of art, which actually is exactly what they are. Highly erotic art, to be sure.

Women who may previously despaired at the thought of tan lines on their boobs need to have another look at them, with a fresh outlook. Many women have remarked to me that they actually find their breasts "boring," as they are so acclimated to seeing the same bosom every morning while staring at the mirror. What needs to be realized is that they are only "boring" because they are familiar. What is needed is a process of defamiliarization, to help us understand the marvelous aesthetics of the breast.

In the early part of the twentieth century, a literary movement arose in Europe to address the issue of how writers should describe any commonplace human experience that seemed dull. They developed a literary technique, referred to as "formalism," which sought to employ language that rendered even the seemingly mundane as extraordinary. Similarly with breasts, women need to also defamiliarize themselves with their own ingrained notions of how they think men visually perceive their breasts.

I encourage women to get truly acquainted with their own breasts as artistic expressions of their sensual personalities. Look patiently into your mirror, explore the tan lines and see how the visual dichotomy defines its own artistic composition. Comprehend how the different shades of skin tone create an aesthetic jigsaw puzzle, and perhaps you will begin to understand why men 's eyes become hypnotic in their presence. The clash of colors adds immense visual complexity to breasts, which translates into visual eroticism.

Without waxing too poetic, I see much allegory in the color cornucopia of the female bosom. The frequent encounter of light and darker skin on breasts is suggestive to me of the Ying and Yang of life. It reflects sensuality that is both overt and subdued. They manifest the erotic in both its sublime and intense formulations. Like an artist's canvas, they delight the senses of the man who is fortunate enough to observe your bosom, leaving your partner both humbled and excited.

Color shades and differentiation is only one part of the aesthetic tapestry, which are the female breasts. The other aspect involves texture. The primary expression of texture on a woman's breast is found in her nipple.

Viewed on a purely visual-erotic basis, nipples are the single most important aesthetic element on the female breast. Ultimately, it is the nipples that make the breast. They define the sensuality and erotic character of boobs far more than size or even shape, at least when viewed from the mind and libido of males. They are not only intense in their color; their tactile nature creates complex ranges of texture that enrapture the eyes of males.

Let us first explore the anatomy of the nipples. They are typically located near the center of the breasts, surrounded by an area of sensitive, pigmented skin known as the areola. Because the nipple and areola are erotic receptors, they should be viewed as sex organs in the same manner as the clitoris. Sexual arousal, temperature change (especially exposure to cold weather or water), breastfeeding and stimulation can cause the nipples to become erect. Women experiencing pregnancy or nursing a baby may see their nipple size become significantly enhanced

Besides milk ducts, nipples contain small non-striated muscle cells called myocytes, arranged cylindrically within the nipple. It is these that are responsible for nipple erections, facilitated by the numerous nerve endings that converge within this small space on a woman's body, overwhelming it with sensitivity.

The anatomical background I just described is important to take note of, as it explains what is unique about nipples. In our discussions, I emphasized how important motion is in generating sensual erotic impressions from the female body. Due to the highly sensitive nature of female nipples, for the reasons just described, they have literally minds of their own. Their turbocharged sensitivity will cause them and the surrounding areola to become rigid, compressed and erect, or recessed and spread out, often in what appears to be a random sequence. Sometimes the two nipples will behave as twins, while at other times they will respond with radical asymmetry. They are so independent minded they often advertise their erections through layers of bra and blouse. When viewing clothed females, the visible erection of a nipple is far more erotically alluring to men than a C or D cup bust line.

As aesthetic accoutrements to the breasts, nipples are inherently complex. Their color frequently clashes with that of the surrounding breast tissue, which is already often dichotomized by tan lines. Increased blood flow during nipple erections can darken the color of the nipple, while radically altering the texture and appearance of the surrounding areola.

Nipples that are erect, enveloped by a rigid areola with a corrugated texture, are intensely erotic to the male visual senses. Men almost feel as though they are witnessing a divine creation when their eyes reveal so powerful a visual sign of feminine power. In a state of erection, nipples appear to men to pulsate with erotic energy.

As with the breast, size is not an important element in determining the sensuality of the nipple. The immense diversity of nipple sizes, along with colors and textures, are further indications of the unique erotic personality of the woman who possesses them. It is truly miraculous that nipples look as different as do women's faces. This applies equally when they are in a relaxed state, as well as being excitably erect.

Sheldon Filger

I have always found it fascinating to observe how nipples can undergo so radical a visual metamorphosis. One moment they lie passively on a woman's breast, resembling an abstract interpretation of an omelet. Moments later, the tissue is rigid and prickly, with the nipple transformed into the nose cone of an erotic guided missile.

At nighttime, or when the lights are otherwise dimmed, the erect nipple can often be viewed while the remainder of the breast tissue is obscured in nocturnal darkness. Women should experience the visual power of their nipples under varying lighting conditions. They should internalize its sensual majesty in near-dark conditions, as they proclaim the sensual presence of an erotic woman even with just a glimmer of illumination.

Nipples are a visual paradox. They can evoke mystical power, yet bask in fragility. Their symphony of constant transformation adds to their surreal character. Often, nipples that appear relaxed, with the areola spread out in a wide circle, are every bit as erotic to men as when they are seen in a fully erect state. It is their wild unpredictability that enhances their innate power to tantalize the senses and electrify the male libido.

In addition to shape and texture, the nipple and its surrounding areola are also characterized by coloring that it distinct from the surrounding breast tissue. It is the discernible coloring of the nipple that

48

infuses the breast with the unique personality of its owner. Sometimes, the nipple's color is a shade more pigmented than the skin on the breast. Often, it is seductively darker. In particular, dark-skinned Latino women, and ladies of African heritage, have exceptionally dark nipples, often with an alluring purple tint. On many Caucasian women, I have observed breasts that were milky-white, enveloping a large, silver dollar shaped nipple and areola of reddish-pink hue, situated like a Zen-paradox on the woman's bosom. The aspect of color, combined with texture and tonality, driven by its innate sensitivity, gives the female nipple a mystical dimension within the pantheon of female sensuality.

Having explored the breast from a number of different angles, let's have a look at them in their horizontal state, as the woman lies down on her bed or couch. We again see the radical transformation of line, as the bosom takes hold of a radically different shape. As the sensuous woman lies on her back, gravity stretches her breasts sideways. With large breasts, they will tend to form a pan-like configuration, with the nipples seated seductively in the center of the pan. With smaller breasts, the skin tightens athletically around the chest wall, often with an erect nipple standing gloriously, as though a statue mounted on its plinth. In every case, we again witness a wonderfully complex realignment of shape and texture, further arousing the attention of lovers.

Remaining in her horizontal state, the breast lines again undergo change and transformation, as the sensuous woman stretches her arms behind her neck, exposing the muscles that run underneath her armpits, connecting the breast ligaments. Particularly with dimmed lights, an erotic tapestry consisting of shadows, breast lines, erect nipples and strained muscles is formed, overwhelming the male observer with raw sensuality.

The perspective I have just shared has hopefully illuminated readers on the subject of breast aesthetics. Far more than a fatty lump of tissue, they are infinitely complex visual compositions, comprising myriad visual details that relate to each other in mysterious ways, while constantly altering their visual appearance. Women who understand both the visual complexity of their breasts, and their changing visual state based on body and arm movements, have an inherently more complete understanding of what is visually erotic about their boobs.

From every angle, every conceivable perspective and from all possible observation points, breasts are indeed the hidden second face of an erotic woman, as alluded to earlier. Unveiled, they exude more emotion than a woman's eyes, more power then her arms and legs. They are inherently more complex than even her sensuous mouth and nose, and more suggestive of her feminine nature and innate erotic power. Breasts are beautiful, bold, sexy, and mysteriously infused with erotic power.

THE MAGICAL MYSTERY
FEMALE TORSO TOUR

I adore the female torso. There is no creation in all of nature that is so artfully and seductively erotic. When photographing a nude subject, I can devote hours to an intense visual study of her torso. Indeed, torso studies are among the most popular forms of fine art nude photography.

In anatomical and artistic terms, the torso is the physical structure of the body that extends from the pelvis to the neck. When artists refer to torso studies, they almost invariably are focused on a frontal view of the torso. It is therefore several key visual features, observed frontally, that define the aesthetics of the female torso.

Torsos are intricate artistic compositions. Visually, breasts, navel, shoulders, hips and pelvis dominate them. However, there are many other elements that make a discernible visual impression. Diaphragms are highly prominent, as the bones of the rib cage mark the woman's skin with their indentations. Abdominal muscles can be highly sensual in their perceived tightness, while the softness of the pelvis clashing with a tuft of pubic hair adds its own aesthetic commentary. The roundness of shoulders and prominence of collarbone juxtapose with the raw sensuality of breasts. Uplifted arms expose the tendons of the shaved underarms, as they form an erotic connectivity with stretched breasts.

My overview of the torso hints at the complexity of this visual masterpiece of nature. Virtually every woman I have observed nude with weight proportionate to her height has revealed a torso of deep sensual beauty. The torso also reveals an underlying truth about erotic female bodies: they are always more than the sum of their parts.

Breasts viewed alone are, to some degree, abstractions. However, when they are seen on the same canvas with hips and navel, stretched abdomen and well-defined diaphragm, they become immensely more interesting for the aroused observer. As a holistic composition, the female torso is an individual masterpiece, celebrating the unique erotic beauty of women. It is the cascade of interconnecting bodily "terrain features" that captures a man's interest far more than a single set of "body parts." As stated before, the truly erotic female body is far more than the sum of her parts. When assembled together on the canvas that is the torso, they create a transcending magic.

Think of how uninteresting a symphony orchestra would be if one could only hear one instrument at a time. However, when the conductor exposes the audience in the concert hall to the glorious symbiosis of all the instruments working in harmony, aural magic is the result. In a similar way, observing and perceiving the female torso as a transcending whole results in erotic magic that is utterly breathtaking.

Women, in my experience, often do not have a full realization of the immense eroticism that animates their torso. Crass cultural stereotypes have inhibited the ability of men and women to fully appreciate the seductive beauty and pulsating eroticism that exists in the body of a woman from her hips to her neck. Too many women are left in a wasteland of confusion, obsessively focused on the their breasts, without realizing that boobs are only a particular feature of the torso tapestry. Breasts are far more erotic when viewed as part of a collage of belly button and hips, and other visual features.

In order for women to have a more complete comprehension of their unique erotic physicality, they should come to a higher level of understanding of the aesthetic characteristics of their individual torso. It's time to again visit the mirror, and experience an eye-opening and life-transforming realization.

The torso should be viewed both topless and nude, as each situation brings forth a distinct visual and erotic impression of the torso. When observing the mirror while wearing panties, the torso is exposed from the hips to the neck, with the pubic area concealed. Typically, women look sternly and critically at their breasts, unmindful that it is the whole torso which fascinates men and teases their libidos. To help women build that broader understanding of what is beautiful and sensual about their upper bodies, they should look first at their breasts, then broaden their perspective, incorporating other parts of the torso into their field of vision, along with their bosom. Begin with the breasts, collarbone and shoulders, imagining a frame enveloping those areas, as though you were studying an erotic painting. Now, let's shift that frame, so that it encompasses the lower part of the breasts including the nipples, the area between the lower portion of the diaphragm, navel and surrounding abdomen. These two views should demonstrate the sensual variations that exists within the frontal torso, and how much more erotic interest is evoked when different features are viewed interacting, as opposed to being observed separately.

Torsos are moving canvases, as stretching of the arms obscures the shoulders, lifts the breasts and exposes the underarms. In addition, the skin covering the diaphragm is tightened, bringing into sharp relief the bone structure of the rib cage. Lowering the arms, tilting at the hips and raising a shoulder projects a radical transformation, as the torso appears asymmetrical, tantalizing the observer.

Women should study their torsos as artistic compositions, observing with intensity how visual impressions are transformed by numerous manipulations. Observe your torso frontally, obliquely and in profile, while raising and lowering the arms. Shoulders are very sensuous on a woman's body, so soft in appearance, yet structurally strong. However, stretching your arms to the ceiling will hide the shoulders from view, while your armpits are visually exposed. Many women do not realize that their shaved underarms are highly erotic to the vast majority of male observers. There is, after all, a reason why so many men are enchanted with sleeveless dresses, beyond enjoying the sight of a bare arm and elbow and shoulder. This is an opportune moment to delve into the aspect of "pits."

Heterosexual men are highly aroused at the sight of an exposed female nipple, while being totally uninterested when viewing the male counterpart. The same is true with female underarms. Provided they are shaved, which creates an obvious distinction with men, female armpits are viewed by the aroused observer as connected with the eroticism of the breasts. At a cocktail party, for example, when the woman in the sleeveless dress raises her glass of wine in a toast, thus exposing the delicate and seductive tissues of her underarm, the gentlemen viewing her become visually intrigued. What often goes through their minds is that they are getting a sneak preview at the ligaments that connect like a freight train to the lady's boob, reinforcing the impression that they are observing a sensuous woman with a highly erotic body.

When women have exposed their torsos to the visual acuity of their partners, they should "exercise" their torsos by frequently alternating hand and arm movements, periodically exposing the underarms. A very sensuous position is to intermittently place both your hands behind your neck, while looking seductively at your partner. The torso perspective of exposed underarms connected to stretched and lifted breasts, navel and diaphragm is inherently erotic.

Another perspective on the torso that I find highly sensuous is with arms gently folded below the breasts. In this perspective, the shoulders stand out, with the collarbone acting as a bridge. The roundness of the shoulders juxtaposes with the curvature of the breasts, and a myriad of contrasting shapes, tones and textures situated on the upper part of the torso. The observer becomes mesmerized, overwhelmed with visual complexity of a highly erotic nature.

Navels are a key terrain feature on the torso relief map, being almost as evocative as breasts. They come in a vast array of shapes, adding a highly distinguishing aspect of personality to the torso. The navel is far more erotically potent, however, when viewed in unison with the breasts, rather than as a "stand alone" feature. As with underarms, the heterosexual male finds female belly buttons far more visually interesting and intriguing than those encountered on his own gender. Women viewing their torsos in the mirror should look keenly at their navels, than understand how they contribute to an erotic composition when the abdomen, diaphragm, breasts and shoulders are also within the field of vision.

The above commentary refers specifically to the topless female body. Now, let us have a look at the torso of a nude woman. Once the panties are off, things change in a very radical way.

A completely nude torso is visually far more complex. The pelvis is exposed, hipbones are distinguishable and the pubic hair is unveiled. All these elements add an intricate menu of new colors, textures and tonalities to our even more enhanced erotic canvas.

Below the navel, the most erotic visual element on a woman's body is her pubic hair. Yet this is an area of controversy, as a growing number of women are choosing to eliminate this indelible sign of bodily womanhood. The means used are often very burdensome, including bikini waxes and electrolysis. All this anguish to suppress the female "bush" begs the question: why are so many women choosing this direction?

I believe that there are two principal reasons why many women are having their pubes suppressed. One involves a false notion that men prefer "shaved" women. The other explanation I believe is even more basic; simple peer pressure.

Many college and university students that pose for me are electing to go the wax route. What I hear from them is that this is a fad among many younger women, a sort of body modification right of passage. Younger women view getting a bikini wax in almost the same way as they view tattoos and body piercings. For some reason, this fad has jumped the generations, as now more middle-aged professional women are choosing to go through life bereft of their "short and curlies."

There is a very dark side to this fad, which has distorted the perception of many women on this indispensable signature of their womanly eroticism. Contrary to the perception of many women, the great majority of men find pubic hair very erotic. This is not to say there are not men who prefer this fad, however, this is where the nocturnal aspect emerges.

The men who have a preference for viewing women with bare skin "under there" tend to be the type of men that enjoy pornography. Virtually every pornographic image of women, be they on DVD, video or in so-called "adult "magazines, present them with shaved pubic areas. This is a view that many in the fine arts community who photograph or paint nude women concur on. That a preference defined by pornographic vulgarity has entered mainstream practices is a deeply disturbing trend. It must be understood that nude women bereft of their "fuzz" may suggest to some purveyors of pornography a degraded female who has reverted to a pre-pubescent stage in her life. I find nothing that is the least erotic in this form of imagery. Pornography is the very antithesis of wholesome eroticism, and women can take a stand by defying a trendy fad that has its roots in a conception of women's bodies that is exploitive and degrading.

So, men with vulgar tastes excluded, heterosexual men find pubic hair on a woman's body to be one of the most erotically exciting aspects of her physicality. There are probably a host of explanations for the intense level of male fascination with female fuzz. Undoubtedly, its proximity to and overlapping of the vulva creates a level of visual erotic tension that is catalytic to the male libido. There are other reasons as well, I believe, that relate directly to the sensual aesthetics of a woman's torso.

On the torso of a nude woman, the pubic hair represents an intense visual element that clashes with the surrounding skin tone, injecting an intriguing level of complexity into the aroused viewer's field of vision. This complexity is further exacerbated by the frequently encountered pale tonality of the pelvis.

In our earlier chapter on breasts, I noted that tan lines are frequently featured on female bosoms, and for similar reasons the same phenomenon is encountered when the pelvis is exposed to scrutiny. Especially on well-tanned Caucasian women, the sight of a deliciously brown navel and abdomen tantalizes our observer, contrasting with the delicate whiteness of the pelvis, which sharply contrasts with the rich darkness of a swath of pubic hair. To the excited observing male, the impression is not that dissimilar to an earlier example sighted, where the nipples on a tanned woman with tan lines on her breasts add immeasurably to her visual complexity.

Our point, therefore, is that pubic hair forms an integral aesthetic function in defining what is erotic with the torsos of women. Remove this pivotal terrain feature from the female body, and a seemingly inexplicable void is created. It would be as though one was observing Leonardo da Vinci's *Mona Lisa* at the Louvre, with the smile expunged from her face.

The often stark juxtaposition that dark pubic hair imposes with the surrounding pale tissue attracts the gaze of an aroused observer, compelling his eyes to survey that highly erotic lower portion of the female torso. I believe that nature intended pubic hair for, among other reasons, functioning as an erotic "neon sign" that draws the visual attention of the woman's beloved towards the most mysterious and intriguingly erotic portion of her physical body.

The pubic hair of women is utterly fascinating. As with breasts, they come in many variations. There are literally an infinity of possibilities, in terms of not only color, but also overall shape, texture and density. Most female bushes I have observed form a triangular patch on the lower abdomen, but often they are seen in the form of a line, as though the individual hairs were soldiers lined up in parade formation. They can be thick or sparse in their density. Often, the texture is different, in a subtle way, from the hair on the head. With brunettes, the color of the pubes will invariably match what is on the head. Men are often mesmerized by these two collections of hair on a woman's body, apparently having the same color but differing so vastly in shape, texture and "style."

When it comes to blondes and redheads, things get really interesting. In the vast majority of women with blond or reddish hair, their pubic hair will not match the color of what is on their heads. The difference can sometimes be rather subtle, just a shade darker. Often, the pubic hair is much darker. In some cases, a woman with naturally platinum blond hair will reveal a pubic bush that is so intensely and mysteriously dark, it gives the appearance of belonging to a brunette and not a blond.

Women should view their pubic hair as an essential erotic accouterment to their sensual bodies. Especially when men visually study the nude torsos of their beloved, the visual complexity that pubic hair injects into the erotic panorama unleashes a potent dose of sensual excitability.

Women have long recognized how many men covet viewing their exposed breasts. However, it is clearly apparent that most women are unmindful of how similar is the male desire to peer at their exposed bush. There was a time, however, when the erotic power of female pubic hair was more recognizable in the general culture.

For much of human artistic history, the female bush was "verboten," except within the confines of the marital bed. Even in the Middle Ages, it was common to view naked breasts in recognized great works of art. However, until the Impressionist and post-Impressionist period, artists painting nude female subjects practiced their own form of the bikini waxing technique. Female pubic hair was viewed as so potent in its eroticism, it had to be suppressed outside the bounds of marital intimacy.

Fine art photographers, from the earliest period, recognized the aesthetic and erotic power of a woman's pubic hair, and were not at all hesitant in celebrating it in their portraitures. The great early twentieth century photographer Alfred Steiglitz created a number of magnificent nude studies of his famous artist-wife, Georgia O'Keefe. There is one image of her, in particular, where O'Keefe's dense and mysteriously seductive patch of black pubic hair dominates everything else on her body, including her intricate breasts and sternly complex face.

Outside the fine art world, however, different standards pervade. The proliferation of men's magazines revealed a cornucopia of exposed breasts, but not event a hint of hair growing on the female pelvis. It was not until the 1970s that leading men's magazines in the United States began exposing "bush" in their nude pictorial layouts, in an exaggerated flush of competitiveness that was dubbed the "pubic wars" by pundits. However, by the 1990s, even seemingly "mainstream" men's magazines began replicating the norm established in the pornographic world for what is often derisively referred to as "shaved pussy."

An erotic woman's nude torso will have hair in its crotch. Anything to the contrary mimics the prepubescent body of an underdeveloped girl, no matter how aesthetically pleasing all the other terrain features are to the human eye. A woman's torso without a sensuous bush is like a mountain without a summit, or a bird without its feathers. Whenever I view a woman's torso bereft of pubic hair, my immediate impression is one of incongruity, that something vital for visual completeness has been inexplicably expunged.

I can recall fragile female torsos I have photographed, so petite the hips were not discernible, with very small breasts. Yet, these female bodies still appeared robustly feminine and intensely erotic precisely because they were outfitted with a proud patch of erotic bush. Pubic hair is an essential optical component to perceiving the female form as holistically erotic. All that women need to do is absolutely nothing. No need for the extravagant care that coiffures on the head require. It is maintenance free. Ignore the fad of waxing and electrolysis, and as a woman you will preserve an essential erotic component to your sexy body, delighting the vast majority of men that are stimulated and aroused by the short and curlies that sprout in that mysterious patch, located in the lower portion of the erotic woman's torso.

As we are discussing the lower portion of the female torso, let us explore the curvy issue of hips. As with breasts, the curvature of hips is quintessentially feminine and inherently sexy. Some curves are rather mild, while others appear highly eccentric. There are numerous variations, each aligned with the unique beauty of the woman on whose lower torso they proudly add their own special definition. Hips also frame the delicate-looking skin tone of the lower abdomen and the navel. In a special sense, they are the boundaries that define the lower regions of the erotic torso.

Observed frontally, curved hips framing the blatantly erotic lower abdomen and navel possess an almost surreal quality. When the woman tilts at her waste, forcing the hips to become asymmetrical, they form an intriguing juxtaposition with the diaphragm, which also contributes its own form of asymmetrical composition.

When the hips are viewed obliquely, the profile of the buttocks also comes into view. As the curvature of the female butt is suggestive to men of the curved lines of the breast, their visual contrast with the hips provides for an enticing visual erotic symphony for the man fortunate enough to be observing.

The curved lines of hips and breasts also interact in the erotic composition that is the female torso. They also are highly suggestive of an inherent aesthetic truth: the erotic female torso is largely defined by curvatures. Curved breasts envelop round nipples. Curved hips frame the soft abdomen that encloses a round belly button. Even the bones protruding from the diaphragm add a phalanx of subtle curves to the body's skin. Twisting obliquely hints at the breast-like roundness of the buttocks.

My description above reveals the intensely complex aesthetics of the female torso, and how it is the visual interaction of terrain features and their often curved geometry that are essential in defining the body of a woman as artfully erotic. There is also a predominating proportionality to the aesthetics of female eroticism that reveals itself in the torso. Yet, despite the fundamentals that underline an erotic torso, there remain an incalculable range of visual interpretations in how this majestic female tapestry called the torso can be constructed and arranged. In other words, while each female torso has a menu of aesthetic elements that interact with each other, it is also by definition totally unique. Every woman possesses a torso that is her own, shared by no other female on the planet. Her torso is as unique as her own face. Yet, in a special and seductive way, it is even more visually complex.

What is it about the female torso that is so complicated? In a word, it is eroticism. As the torso is seen as a holistic composition celebrating the unique beauty and sensuality of its unique possessor, it exudes a level of unrestrained eroticism that is so intense, it captures the taste of the most exotic and seasoned of foods. Just as one looks at a table with an array of mouth-watering food with the senses aroused and mouth salivating, so one feels the sensual impact of the exposed female torso. The visual acquisition of a man's beloved, with her torso unveiled in its delectable nakedness, arouses his senses, overloads his emotions and energizes his libido.

The female torso is indeed magical and mysterious in its representation of the ultimate beauty. It mystically transcends all its marvelous aesthetic visual parts and terrain features, leaving the humbled and deeply moved male observer utterly in awe, as he confronts a composition of unrestrained erotic beauty.

VAGINA DIALOGUES

There is no part of the female anatomy that arouses such intense discomfort and fear in men and woman as the vulva. Not so much in terms of intercourse, with the lights out and eyes closed. What I am specifically referring to is viewing the female genitalia as an erotic bodily composition. Let's face it, while men and women are quite comfortable with the notion that a lover should devote time to visually examining breasts, we do not have the same level of cultural acceptance when it comes to observing the vulva with studious curiosity.

And yet, the vulva does exist, and though it is the most hidden aspect of the female erotic body, it does have its aesthetic qualities and unique physical sensuality. What has been lacking is any frank discourse on the erotic aesthetics of a woman's genitalia.

It is quite understandable that our culture has not been universally receptive to the idea of viewing the vagina and surrounding tissues as beautiful erotic features of a woman. From the dawn of time, it has been a "no go" area for most of humanity's artistic history. Even in our contemporary times, fine art photographers, as a general rule, avoid creating images of the female genitals. I myself have followed this rule, as the concept of candidly photographing the most intimate and sexual part of the female anatomy somehow has struck me as evasive beyond the legitimate boundaries of artistic exploration. This view is reinforced by the fact that almost all photographic images of the female sex organs are found on the pages and in the videos of the most degrading forms of pernicious pornography.

And yet, if not in the public forum of art and culture, at least in the intimate confines of the bedroom shared by a couple in love, the vulva region is an aesthetic masterpiece of erotic complexity that should be studied with fascination and observed keenly, rather than being dealt with solely as an unseen object of female sexuality.

In the world of art, the only examples of tasteful and intriguing representation of the vulva that I have encountered were paintings, and not photographs. All these painting were the work of one artist who, not surprisingly, was a woman. Her name was Georgia O'Keefe, referred to in an earlier chapter in connection with posing nude for her photographer husband, Alfred Steiglitz.

O'Keefe painted a series of pastels that, on the surface, appeared to be unusually sensual and delicate representations of flower petals. Yet, to most women viewing her work, these were not mere flowers, but rather a none-too-subtle allegorical representation of the female genitalia. There is a story told of an unsuspecting man who bought one of O'Keefe's "flower" paintings to decorate the home he shared with his wife. When his wife came home and encountered the O'Keefe painting, she yelled indignantly at her husband, "get that vagina off my wall!"

With unusual perception and artistic courage, O'Keefe had the insight to understand the erotic aesthetics of the female genitalia, and relate that understanding through sensual imagery. The softness, delicacy and fragile yet powerful eroticism of flower petals are truly an insightful representation of the vulva region. What Georgia O'Keefe also taught us through her series of erotic flower portraitures is that the vulva, as with other erotic elements of the female body, is not a one-size-fits-all cookie cutter sexual organ, but rather an inherently complex erotic structure that is uniquely represented on each woman.

It is incumbent upon women and their partners to visually explore their vaginas and surrounding tissues, with patience and excitement, as though viewing a Georgia O'Keefe flower painting at the O'Keefe Museum in Santa Fe, New Mexico. This is best done as a couple, with the woman's partner holding a mirror that will reflect an exposed view of the vulva in a lying posture with legs fully extended.

In viewing the exposed vulva fully extended, the aesthetic importance of pubic hair becomes emphatic in its visual impact. Typically, a tuft of thick pubic hair emerges a few millimeters above the clitoris, covering the mons veneris, a cushion of fat over the pubic bone. In many women, especially brunettes, the hair extends along the posterior of the vulva, sometimes terminating at the anus. This "moustache" of pubic hair creates an erotic frame, focusing the visual field on the alluring tissues it encompasses, while providing a rich contrast in colors. The delicate pink tones of the tissues incorporated within the vulva juxtapose with the surrounding pubic hair, suggestive of erotic tension and incomprehensible mystery.

The lips of the vagina are formed below the pubic bone as two parallel folds of tissues, which surround the vaginal opening, meatus and clitoris. The larger fold, the labia majora or large lips, is fatty tissue covered with pubic hair. Near the vaginal opening it has an appearance that is moist and delicate.

Within the labia majora is a smaller fold of tissue, the labia minora or small lips. It is reddish in color, and can alter its appearance due to its sensitivity. A woman experiencing arousal will have her labia minora become somewhat erect. The top folds of the labia minora join together to form the prepuce, which partially covers the clitoris as if it were the sheath of a sword.

The variations in the appearance of the labia minora are vast. The height on average can range from 10 to 16 millimeters. On brunettes, they tend to be highly pigmented. On the posterior, they can fade away into surrounding tissue, fuse together or even extend to near the anus. The visual interaction of the inner lips with the clitoris and prepuce as well as the outer lips and surrounding pubic hair is fascinating in the extreme. Rather than a gynecological phenomenon, what is observed is the most intimate erotic composition on a woman's body. An intensely personal erotic composition, to be sure.

The clitoris is the most sexually sensitive organ on the female body. Visually, viewed as a solitary object, its appearance is largely abstract due to its small size and significant concealment by the prepuce, even when in a state of erection. However, when a woman and her partner view her clitoris as part of an erotic collage, which includes the delicate tissues and folds of the major and minor lips and surrounding pubic hair, the most intimate erotic imagery is revealed. It is the woman's unique biological representation of Georgia O'Keefe's erotic flower pastels.

Only an erotic woman and her partner will be privileged to observe her own unique vulva composition. Having the courage and audacity to view, together, this "hidden" visual frontier at the very summit of eroticism is an indispensable marker on the voyage to realizing the full nature of your erotic body.

When the vulva is first viewed in a lying posture, with legs extended, a very open perspective is revealed. Understood for its aesthetic and sensual beauty, the external female genitalia are glorious in their triumphal structural complexity.

As with other regions of the female body, different perspectives of the vulva add to our artistic appreciation for this supinely erotic territory. The vulva can be viewed with the position of the legs changed. It can also be observed from behind, with the cheeks of the buttocks stretched apart and peering at a gravitating vulva that appears to float almost magically in front of the anus. Another interesting perspective is viewed when the woman lies sideways, with one leg stretched upwards.

With every different perspective and posture, the complex interaction of major and minor lips and their connecting tissues create an infinite variation of visual impressions, each one uniquely and powerfully erotic. Both man and woman, working collaboratively, should explore all the physical variations and their visual impressions with passionate interest. There is probably no abstract-expressionist painter that has ever created as complex and inexplicable an artistic composition as can be found between the thighs of an erotic woman.

When contemplating the attributes that contribute to our understanding of the aesthetics of an erotic woman's body, the vagina cannot be left out of the conversation. Discussing the sensual beauty of the vulva is far more difficult in our culture than is the case with breasts. And yet, far more than breasts, it is the vagina that is the single most important organ of feminine existence. It is illogical to separate the vagina from a woman's body, notwithstanding the discomfort in treating it as a serious aesthetic accompaniment to the female body's sensual nature.

Beyond the public arena, and solely within the confines of the bedroom, both woman and her intimate partner must see her erotic body as far more than an agglomeration of hips, breast, face, legs and navel. For visual completeness, the erotic body must be seen as having external genitalia. And more than just seen. The vulva should be displayed and celebrated for its sheer erotic complexity, the mysterious nature of its intricate folds, the powerful visual collision of moist tissues, erect structures, hair and secretions.

In her erotic representations of the vulva region and its starkly seductive tissues, Georgia O'Keefe's gutsy artwork celebrated the very essence of womanhood and female eroticism. However, in addition to the allegory of flower petals, there is perhaps another metaphor that is appropriate.

When I reflect on the hidden mystification of the vagina and surrounding region of a woman's body, I think of an enchanted forest. Dark and damp, foreboding at first, a territory initially viewed with considerable caution.

However, when the sunlight is shone on the primeval forest, the enchantment truly comes to life. The bark on the trees becomes inexplicably beautiful. The leaves glisten. Moisture interacts with light, producing imagery that is both incomprehensible yet clearly divinely beautiful.

The vulva represents the enchanted forest of the female body. When it has arisen from the primeval darkness and basks in the illumination of a passionate pair of eyes, both man and woman can rejoice in the ultimate expression of the erotic female body.

74

LEGS, BACK AND BUTT

Considering the salient fact that legs represent most of the length of the human body, it is not surprising that they occupy an important visual role in defining a sensuous woman's physicality. In parallel with men, a woman's legs perform critical functionality. However, as with other aspects that distinguish the female form from its male counterpart, legs are far more than utilitarian components to the female body. They are also objects that beautify a woman, and provide an important aesthetic backdrop for the erotic female body.

Along with a woman's face, hands and arms, her legs are the most commonly viewed aspects of her physical body. High fashion recognizes the beauty and sensuality of a woman's legs. Women's wear frequently is designed to expose as much of the leg as possible, even in a working or professional environment. Stockings designed for women are often transparent, in recognition of the beauty of a woman's exposed legs.

In most of the world's cultures, a woman's legs are not only meant to be seen, but also to exude an appearance of sensual smoothness. In contrast with the hairy legs of men, a woman's pair of legs is meant to glisten in a state of baldness. As with underarms, a sensuous woman shaves her legs religiously, or employs other effective methods of hair removal.

The appearance of a smooth pair of female legs is highly pleasing to men. It connotes a range of impressions, often of a contradictory nature. They suggest strength but also fragility. They are objects of warmth, but also project a surreal quality, especially when juxtaposed with a woman's facial image. They can be viewed statically, yet more often than not are encountered in motion. They suggest the presence of a woman who is desired, and just as easily is indicative of the same woman abandoning the scene.

When the female body is viewed fully nude, the legs add the missing complement that the torso requires for erotic completeness. Add a glistening, smooth yet strong pair of legs to the complex and cur-vaceous tapestry of the torso, and a complete erotic body becomes fully formed. The legs themselves become more sensual in appear-ance when the field of vision includes breast, diaphragm, naval and pubic hair. The collision of erotic bodily features becomes even more intense, inculcating the male observer with the sensual energy that flows through an erotic woman's body.

Legs are far more erotic when viewed as an integral part of the whole erotic body, from different angles and perspectives. When viewing the female body frontally, legs combine with the torso to communicate one range of impressions. Observing the body in profile, the legs connect with the outline of the torso, which is dominated by the curvatures of the breasts, with a seductive bridge formed by the roundness of the buttocks.

When a woman's body is viewed from behind, however, a vastly different range of visual impressions is created that is no less erotic than a frontal perspective.

The nude rear perspective of the female body is often underrated by both men and women for its sensuality. In truth, it is a powerful competitor with the frontal torso view for its naked eroticism. When a woman's bare back, buttocks and legs are exposed for erotic scrutiny, they overwhelm her passionate partner with muscular arousal.

Contrasting the front and back perspectives of the female body, the frontal torso view, in particular, exposes feminine fragility and is overt in its imagery. A woman's back, on the other hand, is somewhat subtler. It is the smoothness and supine nature of the legs, tightness of the buttocks, and elaborately configured state of the back's bone structure that proclaim a more restrained impression of femininity. Yet, a rear perspective is also suggestive of strength and power that is not as visually overt when a frontal perspective is observed. It is that impression of almost statuesque strength that enables the female butt, back and legs to convey a different but no less powerful impression of female bodily eroticism.

Some of the most erotic photographic images of women involve nude perspective of the rear aspect of the female body. There is something tantalizing and irresistible when confronted with the collective imagery of female buttocks, back and legs. There is a delicacy interacting with strength and energy that cannot be encountered on the typical male body when viewed from the same perspective. Both women and men should rejoice that all points of observation of the female body reveal sensuality in abundance.

As the buttocks are right at the center of a rear view of a woman's body, it would be useful to discuss this region from the point of view of erotic aesthetics.

The female buttocks mimic the rounded breasts and severely defined cleavage of a buxom woman. In a sense, they form a second pair of breasts, at least aesthetically, though their shape will almost invariable be distinguishable from the breasts on the upper torso. Their roundness also contrasts with the smooth contours of the shoulders when the arms are at rest, providing the eye with a delectable juxtaposition.

Most men have preference for what is commonly referred to as a "tight ass." While slim buttocks are generally attractive to men, women who are broad beamed need not despair. In many cases, females whose weight is proportionate to their height have body builds that lead to broad shaped butts that do not in any way look obese or unnatural. I have observed many broad beamed female buttocks where the checks were muscular and tight, with no sagging skin from excess fat. Those buttocks were no less sensuous than those of slimmer dimensions.

As with breasts and their relationship to other features of the female torso, a woman's buttocks appear far more sensuous to men when viewed in conjunction with the back and legs. The composition will then appear complete, glowing with erotic imagery.

The roundness of the butt is an intriguing contrast with the myriad of visually complex muscles and bones that proliferate on the backs of women. Taken as a whole, one views the celebratory work of a divinely inspired architect. There appear to be an endless range of complex shapes and geometric abstraction that connect and collide on a unitary canvas. And yet, that is just the beginning. When the woman is in motion, the canvas displays a continuity of changes. Ligaments in the back move and meander as they press against the skin, while bones transform their indentations. Frequently, the buttocks will display its own transformation, often with an ephemeral dimple.

Many women are obsessed with a fear that their buttocks are too "big." This is often paranoia, which can be neutralized through simple visual realization. With the use of multiple mirrors, women need to become better acquainted with their backsides. Rather than focus on the butt, women should view leg, buttocks and back as a single bodily canvas. As mentioned earlier, the butt is the aesthetic bridge than connects back with legs. If the connectivity appears natural, with flowing lines, than this is proof that all is aesthetically in order behind there.

A female nude standing or walking, with back exposed to her partner, shares a severely underutilized perspective of her erotic body. This is no doubt due to the lack of audible male rhetoric with respect to female backs, in stark contradistinction with certain features on the "other side." However, a thousand years of great paintings and more than a century of photography exuding celebratory praise of the feminine back as an objet d'art are more tangible testimony than the silence emanating from the purveyors of mass culture.

Many fashion designers harbor an intuitive understanding of the potent sensuality of the backside of erotic women. This is reflected in classy evening dresses that expose the back down to the emergence of the buttocks, revealing a tiny hint of cleavage. Often, the fabric around the buttocks is so tight, it overtly hints at its seductive shape. Legs are either revealed overtly or covertly, often through sheer or transparent fabrics.

My own view on the sensuality of the female back as it relates to high fashion is that rear-revealing dresses reflect a latent desire to outflank Victorian mores on female nudity. Especially in American society, public display of breasts in a normal setting (as opposed to demeaning settings, as with a "strip club") is suppressed, either socially or legally. We seem to have an intense collective desire to compensate by laying a woman's backside bare from the shoulders to the top of the butt at the most formal of social occasions, including weddings, awards dinners and charity balls.

As delectably sensuous as is the rear view of a standing or walking nude woman, this only hints at the visual possibilities. Some of the most erotic rear views I have observed of a woman's body involved her seated on the floor, weight upon the ass, back bent forward, with the legs bent and twisted in front. The visual impression was of a highly erotic work of modern sculpture. Women and their partners should experiment with variations of this posture. No need to visit a museum of modern art when a living, breathing and exciting masterpiece is seated before you, right on the den floor.

When viewing the female backside, I have always been intrigued by the elaborate collision of lines and angles. In a standing posture, a woman's back is delicately bisected by the indented line that defines her spinal column, traversing a trench from just below the neck to a few inches above the buttocks. When she tilts at her hip to one side, the linearity of that spinal trench becomes seductively curved. One shoulder blade pops out, while the other recedes, with dimples forming on the upper portion of the back, just below the shoulders. Down below, the butt cheeks become somewhat asymmetrical. When the upper part of the legs are added to this visual panorama, on overtly erotic yet mysteriously abstract composition is created.

Another intriguing perspective of the female backside is to view her obliquely, with the full back and a side of the ribcage exposed. One butt cheek is seen frontally, while the other is exposed in profile, replicating the visual experience of observing breasts. The sight of smooth legs extending below the buttocks adds an element of sensual grace as well as power. With this posture, a back bent slightly forward creates a sensually smooth undulation, as the eyes of the beloved scan upward towards the shoulders.

We think of a woman's arms and hands as body parts we only view frontally. Yet, in a backside perspective, a woman does not lack the ability to stretch her arms behind her back. I consider it a highly erotic visual commentary to view a woman's backside, with one arm stretched behind her back, the hand delicately balanced on the waist. The juxtaposition of the hands, with fingers stretched, just above an exposed butt cheek, with her smooth leg extending towards the floor, is a truly breathtaking view for the erotic connoisseur.

Her partner can also appreciate the backside of the erotic woman when she is observed laterally. A nude woman lying down on her stomach provides a unique view of her back, buttocks and legs at rest. As she relaxes, the male observer is anything but relaxed, as he surveys the abstract rear erotica of the woman who intrigues his mind and tantalizes his libido. In this lateral perspective, the woman's partner can view the sensual sculpture her figure represents while she is in repose. He will view her from a profile perspective, than from directly above. The eyes uncover the intricate character of the female back's muscular and skeletal structure. The numerous lines and indentations bring subconscious arousal to her partner. The sloping hills of the buttocks create a dominating terrain feature of pulsating sensual energy. The thighs and calves of the legs are permeated with her essence and energy. When all these visual elements come together as a unitary composition, the result is an immensely powerful erotic portrait.

There is no more sexually seductive companion to the raw sensuality of the female torso with its alluring breasts, abdominal regions and navel than those regions of aesthetic enticement that lie just behind her.

BODY MODIFICATIONS: TATTOOS AND PIERCING

Among the many women who have posed nude for my art studies, an increasing number come with accessories to their bodily skin that nature did not originally intend for them. Tattoos and body piercing abound. Tattoos on every conceivable part of the body. Challenging every preconception. They are joined by a legion of ideas as to where metal piercings can penetrate the body's surface, weaving their own unique visual signature.

In opening this discussion on tattoos and piercing in connection with the female body, I do have a confession to make. I am somewhat of a prude on this issue. My own preference is for a woman's body in as natural a state as possible, unadorned with ink and metal. There was a time when I was highly reluctant to photograph a nude subject with these accoutrements.

My attitude on the aesthetics of tattoos and body piercing has undergone somewhat of an evolution, largely based on sheer necessity. An increasing proportion of the college women that comprise most of my artistic models are tattooed and pierced. If I maintained my reluctance to photograph women who were not in a completely natural state, I would have been left with far few subjects to pose in my studio.

As I gradually accommodated myself to this growing trend among younger women, I also developed a somewhat more enlightened and open-minded view on the subject. As so many contemporary women are enticed by this form of body modification, I will interject my own point of view and ideas on this topical issue. As tattoos and piercing are profound aesthetic complements to the natural female body, it is apropos that I discuss their role in the understanding of what comprises an erotic woman's body.

Body modification art is not a new phenomenon, though its emergence as a right of passage among Western women is a relatively recent development. However, since the dawn of time, humanity from many cultures has experimented with piercing beyond earrings, as well as tattoos, on both men and women.

In African culture, modifications to the body can be of an extreme nature, including mutilation of the skin as a form of decorative art. Indian culture has long celebrated the placement of metal objects on women in areas that were, until recently, not in accord with Western cultural taboos. In particular, Indian women have a long tradition of wearing nose rings.

In Western culture, the fad for tattoos and body piercing seems to have had a multitude of origins. In male culture, elaborate body tattoos have long been popular in sub-cultural and group settings perceived as being highly masculine, including military organizations and prison populations. In the biker world, a similar manifestation has existed, which also involved female participants in that culture. All these elements seem to have been integrated within the coed population of the punk rock community, and from those origins migrated into mainstream culture.

I have observed so many nude bodies of highly intelligent women with an admixture of metal and ink, it leaves me convinced that this is now a significant preference among many young women, irrespective of educational level and socio-economic background. In particular, I have been impressed by the number of graduate students from highly competitive universities who have posed nude for me, revealing a challenging assortment of tattoos and body piercing. Clearly, there is something happening here that goes far beyond fad or simple peer pressure. Prude that I am, I have been forced to open my eyes and my ears, and try to understand this obsession with adding a human touch to what Mother Nature originally intended.

My understanding of the drive many women have for adding ink and metal to their bodies has undergone a metamorphosis. Originally, I was convinced that this was another fad that was driven by peer pressure. However, I now have a fundamentally different perspective on the motivation so many women have for adding decorative body modification to what Mother Nature has provided them.

We men are often not fully cognizant of the powerful social stereotypes that the media and popular culture have inculcated into women as to how their bodies should look. In my mind, many women are attracted to body art as a form of empowerment. It is a way that intelligent and strong-willed young women can "stick it" to the commercialization and objectification of the female body, and add their own aesthetic adaptations in a manner that is pleasing to them.

I have observed a number of intriguing patterns that have demonstrated how many women view piercing and tattoos as a way of making a statement about their own personality and values through a modality that artfully relates to their physicality as women. For example, I have seen unpretentious tattoos that commemorate an important sporting or group event that had a life-transforming outcome. My eyes have surveyed ornate ink patterns that have covered shoulders and buttocks that were personally designed by the woman who possessed that decorated body. I have encountered women that were as comfortable and aligned with their nose rings as with their earrings.

After observing so many women who have been pierced and tattooed, I am not always comfortable with what I see. Some of the body modifications on women I have witnessed strike me as too flamboyant. Yet, more often than not, I end up appreciating the underlying creative energy and personalization that is manifested in the decorative body art I so often gaze upon.

To begin with, I still earnestly believe that a woman can do no wrong by sticking with her natural body. By limiting piercing to her ears, and wearing seductive earrings, with no other form of body modification, a woman's eroticism will seethe through every pore, without any requirement for added ornamentation.

For the many women who do feel the urge to add a personal commentary in ink and steel to their physical form, there are a number of observations I would like to share.

Let's begin with navel rings, probably the most common form of body modification on the female torso. Belly button rings can be a highly erotic complement to the sensual shape and seductive geometric depression that is the female navel. These accoutrements work best when they are complementary, in the same manner that earrings are effective sensual backdrops to the erotic complexity of the ear's shape and inner folds. The unadorned navel of a woman is very sexy, so a navel ring should seek to enhance what is already there, rather than conceal it.

Navel rings that I think are antithetical to eroticism are those that are far too ornate, with precious stones that completely cover up the seductive belly button, transforming it into a mere pad for a jewelry box. Viewing a female torso with a large shiny stone where the navel should be is aesthetically disjointed, and serves to deconstruct the image a man has of a sensual woman's body.

The most aesthetically pleasing navel rings are simple steel artifacts that do not obstruct a clear perception that in the center of the female abdomen lies a sensuous button-shaped natural hole. By being relatively small and not at all blatant, simple steel rings can add a strong dose of feminine enchantment to the lower regions of the female torso. I have noticed in nocturnal surroundings, with low-level illumination present, metallic belly button rings will reflect flickering light, interacting with the shadowing that tends to form on the abdomen in subdued lighting. The visual interaction can be quite exhilarating, as your partner's eyes dance around your stomach, absorbing the complex interaction of shapes, color and texture. Navel rings are a concrete demonstration that metal and torso can go together.

Some feminine piercing, to my way of thinking, is still somewhat hard to comprehend, at least in aesthetic terms. A growing number of college-aged women are electing to have what are referred to as "hood piercing." To the uninitiated, a hood piercing is a pierced clitoris.

I still have not been able to fathom why some women are fascinated with having a thin rod of steel penetrate and hang ornamentally on the most sexually sensitive bodily member present on the female anatomy. There are some young ladies who claim it enhances their sexual satisfaction during orgasm, though I have also heard contrary views from those so pierced.

The issue of hood piercing is surprisingly one that is differentiated. As it turns out, clitoral piercing comes in two fashions: vertical and horizontal. It should be obvious to the reader what the distinctions are.

Whatever the other benefits of hood piercing may be, I am unconvinced of their aesthetic value in enhancing the eroticism of the female body in general, and of her vulva in particular. It just strikes me as too incongruous to belong in that "down under" region of the erotic female body.

The other forms of body piercing that I also have mixed feelings about are nipple rings. Though I have tried hard to understand them as objects of personalization that complement the breasts, somehow, they strike me as more of a comedic than erotic accouterment. Nipple rings transform the fragile sensuous beauty of female breasts into a form of abstract mockery. It is almost as though women with nipple rings view their breasts as torso twins of the ears. If anything, such piecing seems to deconstruct the eroticism of the breasts.

Beyond the aesthetic-erotic argument, nipple rings may be dangerous. There have been cases reported where nipple piercing led to serious infections, with painful consequences to the woman's health. I can't think of anything so antithetical to an erotic body than a practice such as nipple piercing.

Now that we have penetrated the subject of body modification art, let us have a look at tattoos, and how they fit with the erotic female body.

Feminine tattoos are proliferating. I see them on shoulders, backs and arms, on the cheeks of butts, on the abdomen and breasts, on legs and ankles. They can be a small and solitary visual indicator on the female body, or be encountered in abundance, and in large conspicuous sizes. Some are as subtle as a birthmark, while others are loud and flamboyant, screaming of their presence.

Tattoos are intensely personal, reflecting visually on the woman's unique identity. For that reason, I don't think there are a lot of rules that can be laid down on how tattoos best work aesthetically. Not only body type, but also personalities, have much to do with how successful these ink on body statements are. What I offer is more in the way of perspective, than a regimen on inking female bodies. What I believe validates my perspective is that I have seen so many tattooed female bodies, I believe I can offer insights predicated on my own significant experience in understanding how the erotic female body works aesthetically.

In my judgement, what determines if a tattoo "works" aesthetically is if it complements the woman's body, rather than sticking out as an anomaly, clashing with all the other visual characteristics of the natural female form.

The most difficult feature of the female body to tattoo successfully is the breast. In their natural state they are so beautiful and erotic, anything nature did not intend is incongruous. Having both breasts tattooed with a large design transforms those sensual accoutrements into caricatures. However, I have also seen instances where tattooed breasts did work as a complementary composition.

The best examples I have observed with respect to tattooing the breast involved a small, inconspicuous design on women with small boobs. The combination of a tattoo on a small breast seemed to enhance the athletic character of the woman's upper torso, a common feature with small-breasted women. The other visual aspects that worked involved limiting the application of ink to only one breast, and situating the tattoo at least one inch above the nipple. This combination of factors seemed to have a minimal impact on the natural visual elements that infuse the breast with eroticism (see the chapter on breasts) while adding an intriguing morsel of personal customization to the lady's tit.

Though the torso represents a large space of bodily canvas, I do not believe that it is a great viewing hall for tattoos, at least above the navel. Below the navel, I have observed many examples of tattoos around the lower abdomen, hips and upper thighs that I thought were very effective as visual statements and erotic complements to the woman's body. Typically, the best tattoos in those regions are of small or modest size, with feminine designs, such as floral arrangements. Body tattoos generally don't work well as matched pairs, so if you have a purple violet on your upper left thigh, avoid replicating the same design on the other thigh. Tattoo repitition on the female body directly contradicts the intention of adding personalization, by way of body art, to your physicality.

The backs of women are typically more likely to be tattooed than the front. As women can only view these tattoos with great difficulty, or often not at all, such examples of inked body art are clearly intended as a statement meant for interested observers. That being the case, I have never been able to understand very large back tattoos, many having designs that seemed inspired by Gothic comic books. The wondrous and sensual collage of complex lines and shapes on the female back becomes almost totally obfuscated by such noisy tattoos. It is as though the woman has a subconscious desire to render her back into a sheet of canvas suitable for an Andy Warhohl drawing. Such images may get a second glance, however, they lack the proper aesthetics for arousing erotic appreciation.

As with the front torso, I find most back tattoos work best when they are limited to specific areas. Starting at the top of the back, the shoulders are an alluring space for adding tattoos. As with breasts, only one shoulder should be selected for application of ink. Some of the most interesting and complementary tattoos I have viewed have been butterflies. For a reason I cannot convey in words, I find such designs on the rear of a woman's shoulder to be seductive in an intriguing manner. Before applying the ink to your shoulder, however, be thoughtful to its exact placement and angles. Men will view that tattoo from many different perspectives, as your shoulder moves with your arm. It is important that the design works well when it is viewed from different angles, as the shoulder is raised or lowered.

The buttocks are a common space for tattoo art. Unless the "marked" woman makes frequent sojourns to a nude beach, such tattoos will only be viewed in an intimate setting, apart from a doctor's examination room. The butt tattoos should be tasteful and interesting, keeping in mind it is really intended to communicate something about yourself to another person who is viewing your erotic body. As with breasts and shoulders, I recommend selecting only one cheek of the buttocks for application of an inked design. I would avoid suggestions of other living creatures on the buttocks, such as birds or insects. Small, abstract designs are much more compatible with the rounded, sensuous aesthetics of the unadorned butt cheek.

An area of the back that can accommodate larger designs with aesthetic success is the space just below the bottom of the spinal chord, and above the crack between the buttocks that forms its breast-like cleavage. It is a smooth surface, generally, lacking the more complex admixture of bone and muscle that proliferate through most of the backside. In a sense, it appears that this smooth area was made for displaying tattoos. I have seen some very compelling, intricate, well-sized designs, occupying most of that bodily space. More often than not, I have been impressed with what I have observed. For women who are truly creative, and have the desire to design their own customized tattoo, this zone just above the buttocks is heaven sent. It also displays exceptionally well in a multitude of postures, including standing, sitting, and being in motion. This includes bending exercises, running, walking or dancing. It is also alluring to view while lying down, as your partner's fingers stroke and caress your back.

Legs offer many possibilities for creative tattoos. I would avoid the knees, as the idea of placing artwork on a joint makes as much sense as creating a drawing on a door hinge. The upper thigh, lower leg and ankle are spaces on the female body where I have seen creative, erotic women place their own personal stamp of identity with positive affect. In particular, feminine tattoos on the upper thighs can prove to be a highly stimulating companion to an erotic feminine body.

How do tattoos actually work in integrating their presence with all the other natural elements that cohabit as a single, erotic bodily composition? Women who really understand their own sensuous physicality and its erotic character are able to integrate metal and ink with the natural fabric that cumulatively projects intense erotic visual energy.

An example is how metal and tattoo can work together on the lower torso. Especially with women having fair skin, a flat stomach, tight abdominal muscles, well-defined navel and dense, dark pubic hair, interesting possibilities open up. Add a floral design to a hip or upper thigh, and stainless steel ring to the navel, and an even more complex array of visual textures and tones cascades into the eyes of your partner, overwhelming that lucky observer with powerful torrents of eroticism. The flower so close to the pubic hair is suggestive of an erotic garden, while the admixture of metal, belly button and tight, pale skin provides for an intriguing visual contrast, heightening the excitement and exuding a strong dose of mystery, erotica and sexual energy. Art lovers who have libidos will think they have just been to the most erotic art museum on the planet.

One final thought about body piercing and tattooing, and this has nothing to do with eroticism. Women who choose this route should also carefully select the person who is going to pierce and tattoo them. Only highly qualified, professionally trained persons should be trusted. To do otherwise is to invite risks to your health. After all, there is no form of body art worth dying for.

THE EROTIC MIND HAS
AN EROTIC BODY

Contemporary women have never had so many challenges and opportunities. They live in a world that is both a volatile pressure cooker, and a bridge that crosses the chasm of inequality that afflicted all previous generations of women. Among the many paradoxes, one predominating example relates back to the sensuality of her own body. It can be both a curse and a blessing.

The commoditization of women's bodies as sex objects, devoid of a human presence, has been oppressive to a vast number of women on a manner that is probably not comprehended by most men. However, the increased level of education and literacy by women has provided them with the means to liberate both their minds and their bodies from the crass depersonalization that mass culture and corporate tastelessness has inflicted upon the human race. By appreciating the unique sensual nature of their own physicality, women are actually repudiating the faceless pseudo-eroticism of inane marketing and advertising moguls, while enhancing their own self-esteem and bringing more joy into the lives of the men that they love.

Prior chapters of this book have dealt with bodily aesthetics as a means to enable both women and men to understand what contributes to the existence of an erotic woman's body. In this chapter I want to focus on the mind of a woman.

Is the mind actually an erotic part of the female body? No, it is not a specific part, like breasts, tummy and legs. Rather, the mind is the cosmic glue that binds together all the beautiful aspects and features that predominate within every unique woman's physical topography, creating a transcending halo and life force that is the ultimate erotic driver. An erotic body has an erotic mind, and vice versa. In other words, the single most erotic complement to a sensuous woman is her own erotic brain.

Why is the mind of a woman so important to defining her overall physical body as erotic? I think an excellent way of explaining this point is to think of women who are outwardly physically beautiful, and yet, except for the most boorish and superficial of men are not enticing in a sensuous way.

Women who are externally beautiful yet lacking in sensual attractiveness invariably have minds that do not impress. They may do aerobic exercise to keep their waistlines trim, yet devote little or no time to reading a good book, or attending a lecture, so as to keep their minds fit. Minds that are not fully developed in the intellectual sense, that lack the quality of mental curiosity and the capacity for perception and articulation, are unable to fulfill their essential role within the organic whole that is the erotic female body.

A woman who is highly intelligent will have a face that is animated. Her lips move seductively as she expresses an opinion, offers advice or talks passionately. On the other hand, a vapid lady is often speechless, or speaks only with simplistic cliches. Her face, the movement of her eyes, even her body movement lacks the sensuous grace and spontaneous passion of a mentally agile woman.

My message, therefore, to both men and women, is that smart feminine minds are an indispensable transcending factor in providing an ultimate definition of a truly erotic female body.

I find women who can recite poetry, discuss politics, argue creatively or offer insightful advice, combined with a body figure that has weight roughly proportionate to its height, to be quintessentially erotic. Perhaps what is most sensuous about female intellects is their uniqueness. The voice and thinking process that underlines its inflexions and vocabulary are the ultimate aesthetic accompaniment to all the other physical characteristics we have perused. The great majority of men concur with my perspective. Those that don't are not worthy of having the opportunity of encountering your special and unique erotica.

A woman who wishes to have the most erotic body she is capable of possessing needs to be conscious of health, diet, exercise and grooming. Even more important, she must come to understand her own sensuous physicality, and comprehend it as an erotic composition. And, when she has done all that, she must take advantage of every opportunity to expand her mind and intellectual acumen.

An interesting way for a woman to expand her metal prowess while simultaneously building a stronger appreciation of her own sensual physicality is to partake of as many opportunities for artistic and cultural fulfillment as are available. The best part about this voyage is it can be done both as a solitary activity as well as a bonding exercise with your partner.

Women and their partners would benefit immensely from visiting the local art museum. In looking at works of art that impress you as inspirational, try to understand why that is so. Construct in your own words a rationale for why your emotions were moved by the image or statue or example of pottery. Especially in viewing a painting of a woman that strikes you as exceptional, try to comprehend what was it about the woman in the painting that had such a powerful influence on the artist. In reflecting on that question, ask yourself if it was only the beauty of the woman's face that moved the artist, or is the face as painted merely an indication of what the painter felt about the personality and human presence of that woman.

Viewing how women are presented in great works of art is one of the most uplifting ways for women and men to demolish the commercialized stereotypes of what beautiful and interesting women should look like. A powerful explanation for this is that these women appear as human beings with personality and feelings, unlike the plastic images found on television and other media.

There was a time, before the advent of electronic media and crass culture, when exceptional men were truly attracted to exceptional women. Even in the darkness of the Middle Ages, a chivalrous knighthood would compose the most delicate love letters to the women they loved. The literary exchange between man and woman was characterized by a level of amorous sophistication that seems inexplicable to us in the supposedly more advanced twenty-first century. The famous love letters between Abelard and Heloise remain as compelling today as when they were originally written in the twelfth century.

In the early nineteenth century, the renowned pianist Franz Liszt developed a torrid affection for the strikingly beautiful Countess D'Agoult. In one of his letters to his beloved, Liszt wrote:

"My heart overflows with emotion and joy! I do not know what heavenly languor, what infinite pleasure permeates it and burns me up. It is as if I had never loved!!! Tell me whence these uncanny disturbances spring, these inexpressible foretastes of delight, these divine, tremors of love. Oh! All this can only spring from you, sister, angel, woman, Marie! All this can only be, is surely nothing less than a gentle ray streaming from your fiery soul, or else some secret poignant teardrop which you have long since left in my breast."

Liszt was clearly overwhelmed by the eroticism of the Countess. Yet, it is to her "fiery soul" that he looks to as the source of all the tremors of love aroused in so potent a manner as to compel him to write, in describing his life before meeting her, "it is as though I had never loved!"

Like the Countess who infused Liszt with the potency of her erotic love, every woman who aspires to have an erotic body must have it animated by a "fiery soul."

Theologians have their own understanding of what defines a human soul, at least in the celestial world. Here, on planet Earth, soul is that special quality of the human mind that gives it personality, presence and unique emotive identity. The souls of women, meaning their feminine minds, comprise the electrical charge that seers through their nervous system, bringing life force to their physical being. A passionate, fiery soul, illuminated with character and intelligence, is the ultimate indicator of eroticism within the female body.

Intellect is an essential ingredient of an erotic mind. However, beyond sheer brainpower, the sensuous woman has high-esteem, especially in feeling confident about her own body. When a brilliant, interesting woman has been able to internalize how beautiful she is, and comprehends her own erotic physicality, than she has truly broken through all the false myths about female beauty and sensuality. Once a woman has completed that journey, she it truly liberated. A woman who knows intuitively that she possesses a sensuous, magnificent body and has a passionate, erotic presence has attained the ultimate level of female empowerment.

Empowerment is one of the most important objectives for women to achieve. Though many men now have enlightened feelings about sharing power with their female partners in our contemporary world, there still remains much inequality and tension between the two genders. However, by recognizing their own erotic nature and physical beauty, women can attain a level of control in their own lives that will provide much greater personal fulfillment and self-esteem. Most importantly, they will feel that they have full ownership of their own bodies, being able to rejoice in the recognition of their own sensuality.

Knowledge is power. Being of ardent intellect and having insatiable curiosity about acquiring new knowledge, women, with the encouragement of their partners, should follow through on the many suggestions made in this book on how to better understand what contributes to an erotic female body.

Throughout the "Erotic Book," many ideas have been postulated on how women can better understand the aesthetics of female bodily eroticism. If there is an underlying theme to all of my suggestions, it is that women should endeavor to fully comprehend their own unique physicality as a divinely inspired erotic composition. Especially if you have immersed yourself, at least to some degree, into the visual arts, create opportunities to put theory into practice. Visit the wondrously beautiful art museum that exists on your own body. View it ardently, with passion and excitement. Relish the splendor!

I have frequently suggested that women look in their mirrors more often, learning how their own bodies "work" as aesthetic compositions. Personal awareness of the female body by its owner is an essential requirement for internalizing physical eroticism. Beyond mirrors, there is another avenue that may empower women to realize a fuller appreciation of their own innate sensual beauty.

While mirrors are limited in what they can show a woman in terms of an intimate portrayal of her own body, photography does provide that opportunity. With simple digital cameras, a woman can easily photograph her own torso merely by pointing it at her own reflection in the powder room mirror.

When it comes to other regions of the female body, more specifically, a rear perspective, the assistance of a partner or trusted friend should be enlisted. Essentially, the goal should be to create a small portfolio of your own body. Frontal, rear and profile views should be photographed, as well as reclining and lying flat postures. Arms should be placed in a variety of positions, at your side, raised to shoulder height, and stretched upwards. As discussed previously, the multitude of lines and curves visible on the nude female body should be displayed in a variety of states. On frontal views, your face should display a diversity of emotions, from excitement to passivity.

In a quiet moment, a woman who can study images of her own body will acquire a level of knowledge that will prove empowering. Without trying to imagine how others see her, she can learn the fundamental truths about her own physicality. Her own sense of aesthetics will reveal to her the undeniability of her own unique feminine beauty and sensuality.

Partners should encourage the women they love to better understand their own bodies. Allow them the space and time to create a photographic portfolio of their bodies, and become familiar with the images.

As a woman becomes much more intimately acquainted with her own physicality, she will begin to create an inner sense of what aesthetic characteristics she possesses are innately erotic. Her level of understanding as it relates to her own sensuality will be based on a rock-solid and firm foundation: a first-hand knowledge of her own body as an erotic fine art composition.

Given the pressure women have to endure due to societal stereotypes regarding female "looks," and the time they devote to fashion, hair styling, cosmetics, clothes, diet and aerobics classes, it is inexplicable how little time, if any, most women devote to creating a visual understanding of their own bodies. Far too often, they take a quick look in the mirror, write off their physicality as mundane, then move on to other priorities.

Yet, having an inner comprehension of one's own bodily beauty and erotic physicality is of critical importance for women, and the men who love them. There is no firmer foundation to self-esteem, a feeling of confidence and fulfillment than achieving the realization of being a woman who is both physically beautiful and intensely erotic.

Women who have not embarked upon the journey of comprehending their own eroticism tend to resign themselves to an image of being unattractive, physically unexciting, the very antithesis of being a stimulating physical presence. By admitting defeat, the war is already lost.

The men who are passionate about the women they love should do everything possible to facilitate the journey of sensual self-awareness for their partners. It is not enough to tell them they are beautiful. So intense is the pressure by society to accept relegation to being mundane if you are not a woman who conforms to a specific set of looks and physical specifications that are in vogue, that mere words, no matter how heartfelt, are impotent.

It is only when a woman takes that first step towards beginning to learn about her own body as an artistic creation that she will begin to shatter the boundaries of false conformity, thus liberating herself from the prison of aesthetic ignorance. After the initial hesitancy, women begin to internalize their own sense of being beautiful and erotic. They no longer think of their breasts as too small, or being physically too tall, or petite, or just too average. Their own uniqueness becomes known to them as the most sublime and exalted of gifts from Mother Nature.

Having photographed many beautiful women, I am convinced that there is no such thing as the "most beautiful woman in the world."

This is a meaningless statement, intended to stimulate false competitiveness among women towards attaining a mythical visual standard. Almost every woman in good health, having weight roughly correlating with her height, is uniquely beautiful. What women need to focus on is what makes them aesthetically unique, rather than how their looks in general, or specific body parts in particular, compare with those of another woman.

When a woman reviews her own intimate photographic portfolio, she should maintain an attitude of curiosity, as though she is viewing a series of images of an exceptional erotic art composition. While observing those wondrous images for the very first time, do not restrain the feelings of joy, enlightenment and empowering emotion. Then say to yourself, "that erotic body is me!"

Men are deeply aroused by women who are confident about their sensual physicality, and project comfort in being visually erotic. The perception men form of women they see as manifesting erotica comes largely from the energy generated by the female mind. Women with impressive minds, that also exude erotic-self-confidence, communicate to their partners that here stands a woman with a delectably exciting and stimulating erotic body.

For a woman to comprehend her own erotic physicality is actually a far from simple objective. It requires a great deal of sophistication on the part of women in order to attain that level of sensual self-realization. And yet, it is that very level of sophistication that becomes the ultimate icing of the erotic cake which is the female body.

The interaction of a brilliant female mind with her sensual body creates a presence of feminine sophistication. This combination represents the ultimate erotic female body. Men are utterly overwhelmed by such a potent combination of female presence. Mind and matter combine in a powerful and mysterious way, creating a complex web of intelligent voice, animated eye movement reflecting a fiery soul, interwoven with a uniquely and erotically beautiful bodily composition.

The realization by men that they are in the presence of a female body that is so magnificently energized, arouses the most important emotion that can be aroused by feminine beauty; passion. It is the partnership of female mind with female body that lights the flame that sparks the outpouring of passion from one's lover.

Passion is the reward for being of erotic presence. That is why erotic mind and female body are inseparably linked. One without the other negates passion, replacing it with the dull and ordinary.

A female body, no matter how superficially beautiful it may be in appearance, lacks eroticism unless it pulsates with the unique personality and emotion of an intelligent, soulful feminine mind. Otherwise, one is left with a mere effigy. There are actually men who buy inflatable effigies of women, containing the supposedly desirable physical specification. Of course, except in cases of the most grotesque pathology, such men will view those inflatable women as no more than toys.

The objectification of women in our popular culture has encouraged too many men to view women as toys, in almost the same manner as if they were inflatable dolls. Clearly, passion is totally lacking in such human interactions. Ultimately, when women are viewed as toys or objects, this leads to a level of depersonalization that is totally counter to what eroticism is.

A powerful female mind will eliminate the objectification of her own body, and in doing so, give it presence that arouses passion, and restores the integrity of her unique individuality. Eroticism is not so much a question of mind over matter. It is more along the lines that the mind must matter if the body is to be taken seriously.

The union of female physicality and soul are the essential ingredients for creating an erotic cocktail. I reflect on my many photographic art studies of beautiful women. How much more interesting were their physical bodies when I was fully cognizant of the interesting female personality and impressive intelligence that animated those artful human forms.

Thoughtful women are able to build an inner understanding of their physicality that broadens their outlook on what comprises uninhibited eroticism in the female body. Having emphasized the critical dimension of intellect in defining the female body as erotic, we also are saying that eroticism in a subjective interpretation of womanhood.

Subjectivity inserts the personal element, the dynamic of thoughtfulness, into the visual equation. It runs totally counter to objectification of the female body. It establishes the predominant role of female intellectual prowess and personal character in contributing to our understanding of her as a sensual physical presence. The factor of subjectivity establishes the critical role of intellect in comprehending the erotic female body. This applies not only to the woman herself, but also, just as importantly, to her partner.

The example cited above regarding men with a fetish for inflatable female dolls is in a sense a metaphor for the negativity of mass cultural stereotypes and iconography. It is not only the female gender that has been oppressed by this latent crassness in our culture. Men as well have been victimized by the proliferation of vulgarity. In a macabre manner, men are too often told to be wary of women who have strong minds and engaging personalities. Yet, by objectifying women as mere effigies, men too are denied their own fulfillment of savoring the sweet and delectable taste of female erotica.

In the final analysis, the truly erotic female body is not only animated by her own erotic brain, she has a passionate partner who also possesses an erotic mind. When it comes to men, in particular, the final underpinning in the understanding of an erotic female body is his own repudiation of commercialized objectification of women. He is also one who banishes conformity, and has the sense of aesthetic comprehension of the visual and magical presence of the physicality of a truly angelic female soul.

Men who are able to overcome the false barriers to understanding what is truly erotic about the female body are able to look at the seductive curvature of the breasts, unmindful of irrelevancies such as fat content. He views his beloved as a unitary composition, conscious that he is observing the wondrous physicality of a human being who is proudly feminine, brilliant, charismatic and compassionate. He integrates his knowledge of the female personality with his visual appreciation of her physical splendor. This conglomeration of physical and emotional impressions contributes to his reward, a state of arousal and excitability. It is not women per se that excite, but that one woman he feels a bond with that is of magical intensity.

The man who appreciates the eroticism of that special woman in his life is supportive of her desire to understand her own physical sensuality. He encourages her to have a visual comprehension of her own physicality. He is more than eager to participate in that exploration, and is just as at ease when she needs to conduct some of her erotic exploration in moments of privacy.

The partner of a woman with an erotic body is always considerate of her humanity. He never succumbs to objectification. In his manners and speech, he will not ever allow his female partner to be degraded into feeling that she is a mere collection of individual body parts. His consideration of her as a total person, with an artfully beautiful body animated by the remarkable human presence of a feminine mind, is never ambiguous. He knows how to communicate the joy and exultation he is infused with whenever he sets eyes on the erotic physicality of her nude figure. It is the body of the woman who is his passionate interest and terrestrial joy.

The erotic female body, therefore, consists of two minds. One is the mind of that body's possessor, the remarkable and sensuous woman. She has the intellect to make her mark as a personality of interest. This woman also has the intuition to comprehend her own erotic physicality, and rejoice in its latent power.

The other mind that underpins the sensuous woman's erotic body is that of her partner. Together, they have been granted the most visually exciting gift of art that any human being could hope for.

The spectacular and erotic beauty of the sensuous woman is a triumphant creation of an inscrutable universe, for both man and woman to exalt over. The quality of that visualization of beauty, which is both natural and angelic, is so awesome, it defies understanding in words. Yet, as imperfect as words are, the most striking prose in praise of a woman's erotic power may hint at her physicality, but more importantly, illustriously celebrates her fiery soul. The mind of a woman. A woman who possesses the erotic body of infinite desire.

AGELESS SENSUAL BEAUTY

Driven by the insatiable inanities of commercialized media, women are overwhelmed with false rationalizations as to why they do not possess an erotic physicality. They are too fat, albeit five or ten pounds above an anorexic measurement of "ideal" weight for their height. Their breasts are too small. No, they are too big. With skin too darkly tanned, or complexion too pale.

Women are bombarded with messages, subliminal or overt, inculcating them with the belief that they are far too tall, or monstrously petite. Their hair color, eye color, facial expressions, nose shape, chin and hands are just too much of an anomaly to be characteristic of an erotic woman.

Once the erotic woman and her partner overcome the falsification of physical sensuality continuously promulgated by popular mass media, they must still face the most insidious barrier to appreciating the erotic female body. The canard that age is the enemy of sensual beauty.

Popular culture, television and cinema, along with advertising, continually bombard our senses with the iconography of youth as emblematic of sensuality. Many women have a poor image of their own erotic physicality due not so much to physical stereotypes, but far more to the inculcation of a belief that they are too old to be sexy and beautiful.

Of course, men and women who are in the their eighties or nineties, not to mention centenarians, are not likely to win a Mr. and Mrs. Nude World contest. There is a point in the physical development of all living creatures when we all start to fall apart. This is simply a rule of biology.

Nevertheless, with improvements in health, medical care, diet and especially exercise, a greater number of women are retaining the erotic character and sensual beauty of their physical bodies far into middle age. Indeed, the process of aging, during the period of middle age, merely alters or adds character to a woman's erotic body, as opposed to undermining its essential eroticism.

I have observed and photographed nude female bodies covering a wide spectrum of ages. There are a significant number of women who possess bodies during their forties that would put many young college women to shame. Numerous women in their fifties, even in their sixties, who live healthy lifestyles and exercise frequently, possess bodies that are still immersed with potent eroticism.

One factor that again must be emphasized in the fact that a woman with an interesting mind has already laid a firm foundation for being perceived as physically erotic. An increasing number of women in middle and late middle age have reached an impressive peak in their lives, both in terms of professional advancement and intellectual development. These attributes relate directly back to the physical desire they exude in the eyes of men, especially gentlemen of similar background and maturity.

The combination of power, intellectual maturity, charisma and wit, internalized within a physique that reflects the discipline of exercise and diet, is very often highly seductive to many men, and not only those within the same age category. Many younger men find older, mature women who combine physical beauty with intellectual presence to be a volatile erotic cocktail.

Many older women, with terrific physiques, somehow believe that they are no longer physically erotic because their hair has begun to show traces of gray. On a very young woman, gray hair would be an incongruity. However, on mature women, this is a natural and expected evolution of a woman's facial features. On such women, I find a natural gray coiffure to be sexier in appearance than a thinly disguised attempt to camouflage the natural process of aging with dyes.

A woman with graying hair can be a stunning sight in the nude. The visual tension between the gray hair on the sensuous woman's head and dark pubic hair on the lower abdomen, with a sizzling topography of breasts, diaphragm and navel in between, is almost tectonic in its affect on the mature male libido. Grayness in itself has many interesting visual permutations, as the woman's hair transitions over a period of years from a salt-and-pepper appearance to one that is more dominated by gray.

As I have stressed throughout this book, the interaction of many colors and textures are a significant aspect of the female body's erotic character. A head with gray hair is just another new element that adds to the sensual admixture that appears on the female tapestry of physical sensuality.

Some of the most sexually explicit comments I have heard whispered by men about women they felt a strong erotic attraction for involved gray-haired women in senior executive position in the corporate world. Here were women who were tough-minded, ambitious and aggressive, who clearly earned the gray stripes on their coiffures. I remember one such woman, with small breasts and nipples permeating her blouse. Behind her back, I heard the comments of her male subordinates. It was not a promotion that they fantasized over, as her erotic physicality overwhelmed them.

Breasts on older women are every bit as visually alluring to the male libido as they are with younger women. On larger breasts, sagging is a natural development with aging, and is no less sensually attractive. As I stressed in my chapter on breasts, it is their shape more than their size that intrigues the eyes of the woman's partner. A sagging breast line is merely one aesthetic possibility. Others exist, as indicated earlier. The middle-aged woman should view her breast with the same degree of aesthetic mobility as in her youthful years.

Small breasts tend to show relatively little transformation of their aesthetics, at least until the latter stages of life. They may become somewhat wrinkled, which in no way is unattractive to most male observers. I have photographed women who had firm breasts in their fifties, every bit as delectable as they most probably appeared when those women were a mere twenty years of age.

Throughout much of history, women frequently died in childbirth. The average life span for women was typically in the early thirties. However, in our contemporary world, women in the developed nations have attained a life expectancy in excess of 80 years. That being the case, it would be illogical to write off the physical bodies of women who are in their forties, fifties and sixties as being non-sensual. Women in their thirties should be viewed as still youthful, with minimal aesthetic transformation in their bodies from when they were in their twenties. With more opportunities for women to expand their minds while contracting their waistlines through exercise and athletic activity, both men and women have much to look forward to as the erotic female body evolves through life.

A woman's age, except in the latter stages of life, has little if any negative impact on the sensual physicality of the female body. Many women will experience aesthetic changes that offer interesting variation of their sensual physique, which can actually enhance their erotic appearance in the minds of many men.

Ultimately, physical eroticism in a woman's body is timeless. As with false physical stereotypes about feminine visual sensuality, ageism is equally bereft of merit. Forget what Hollywood and Madison Avenue have been conspiring on. Their mythology about a set of physical specifications tied into a narrow (and youthful) age range as the epitome of the feminine sensual physique has as much validity as do most old wives tales. Rather than corporate mass media and the lowest common denominator of the entertainment industry, I trust more in the libido and passion of the men who idolize the female body across the age spectrum, with its infinite variation in height, stature, bust size, hair color and style, skin tones and eye tint.

A woman of fifty should be no less curious to explore her own physicality as is her counterpart at age 18. Be she a college coed, mid career professional woman or stay-at-home mother in her thirties, starting a new career at 45, or embarking on a new relationship at 55, the erotic woman retains her erotic body. She is as mesmerized at the unique sensual artistry that figuratively defines her physicality as is her fortunate partner. Together, as time passes, they delight in the most inspirational gift our remarkable universe has bestowed on the visual senses of the human race. They explore, celebrate and exchange joy in a state of awesome humility. The erotic body of the erotic female personality, amidst the unpredictability and unknown dangers of living, remains the very essence of all that is beautiful in our world.

In closing my commentary on the sensual physicality of womanhood, I reflect on the vast multitude of women who possess erotic bodies. They are the young college freshmen, bursting with feminine enthusiasm and ardent physicality. I think of the graduate students, serious and focused on their academic life, terrific as subjects for my fine art photography, with bodies of wondrous complexity. Then there are the beautiful professional women, secretaries, airline stewardesses, business women and mothers in their thirties, each with an amazing tapestry of artful visual sensuality. I also reflect on the many women I have encountered, older than 40,their faces reflecting wisdom and charisma, with a towering eroticism that manifests from their triumphal nude bodies.

How fortunate we men are to have as our human partners a gender that arouses such ecstasy and elation through the aesthetic miracle that defines their physicality. Praise be to women, whose physical eroticism we cherish to the end of time!

www.ingramcontent.com/pod-product-compliance
Lightning Source LLC
Chambersburg PA
CBHW022004170526
45157CB00003B/1141

* 9 7 8 1 4 2 5 9 2 9 1 7 6 *